STUDENT SOLUTIONS MANUAL
FOR SMITH'S
ESSENTIALS OF
TRIGONOMETRY
Third Edition

CHRISTY CONN
THOMAS JEFFRIES
JOHN MOODY

Brooks/Cole Publishing Company

I**T**P® *An International Thomson Publishing Company*

Pacific Grove • Albany • Belmont • Bonn • Boston • Cincinnati • Detroit • Johannesburg • London
Madrid • Melbourne • Mexico City • New York • Paris • Singapore • Tokyo • Toronto • Washington

Project Development Editor: *Elizabeth Rammel*
Editorial Assistants: *Melissa Duge, Jennifer Wilkinson*
Marketing : *Margaret Parks, Laura Caldwell*
Production: *Dorothy Bell*

Cover Design: *Larry Molmud*
Cover Photo: *Tom Van Sant, Geosphere Project/Photo Researchers*
Printing and Binding: *Patterson Printing*

For more information, contact:

BROOKS/COLE PUBLISHING COMPANY
511 Forest Lodge Road
Pacific Grove, CA 93950
USA

International Thomson Publishing Europe
Berkshire House 168-173
High Holborn
London WC1V 7AA
England

Thomas Nelson Australia
102 Dodds Street
South Melbourne, 3205
Victoria, Australia

Nelson Canada
1120 Birchmount Road
Scarborough, Ontario
Canada M1K 5G4

International Thomson Editores
Seneca 53
Col. Polanco
11560 México, D. F., México

International Thomson Publishing GmbH
Königswinterer Strasse 418
53227 Bonn
Germany

International Thomson Publishing Asia
221 Henderson Road
#05-10 Henderson Building
Singapore 0315

International Thomson Publishing Japan
Hirakawacho Kyowa Building, 3F
2-2-1 Hirakawacho
Chiyoda-ku, Tokyo 102
Japan

Printed in the United States of America

10 9 8 7 6 5 4 3 2 1

ISBN 0-534-35209-X

Table of Contents

CHAPTER 1   RIGHT TRIANGLE TRIGONOMETRY

Problem Set 1.1

1. a. theta   b. alpha   c. phi   d. omega   e. mu

3. a. δ b. ϕ c. θ d. ω e. μ

5. a. 360°   b. 180°

7. a. 90°   b. 135°

9. a. −90°   b. −45°

11. a.   60° is two-thirds of the way through quadrant
   I since each quadrant contains 90°. It is an
   acute angle.

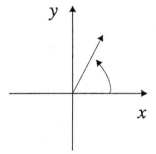

   b.   180° is one-half a whole circle so the angle terminates on the negative x-axis. It is a straight
   angle.

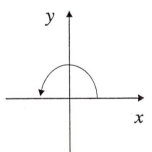

   c.   45° is exactly halfway through quadrant I. It
   is an acute angle.

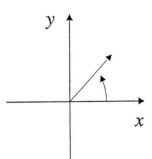

d.   360° is a complete circle.

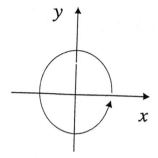

e.   300° is exactly one-third of the way into quadrant IV. It is an obtuse angle.

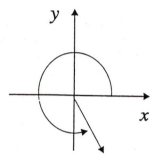

13.   a.   −30° is exactly one-third of the way into quadrant IV moving clockwise from the positive *x*-axis. It is an acute angle.

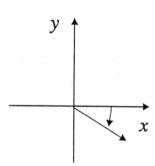

b.   135° is exactly halfway through quadrant II. It is an obtuse angle.

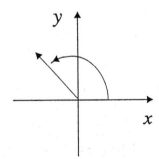

c. −120° moves clockwise from the positive x-axis one-third of the way into quadrant III. It is an obtuse angle.

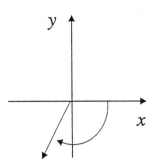

d. −270° moves clockwise from the positive x-axis and terminates on the positive y-axis (at 90°). It is an obtuse angle.

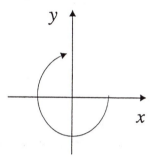

e. −200° moves clockwise from the positive x-axis. It terminates 20° into quadrant II from the negative y-axis. Note that −200° and 160° are coterminal but not equal since they each rotate in different directions. It is an obtuse angle.

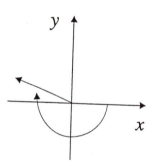

15. a. $65° + \left(\frac{40}{60}\right)° \approx 65.67°$

  b. $146° + \left(\frac{50}{60}\right)° \approx 147.83°$

  c. $85° + \left(\frac{20}{60}\right)° \approx 85.33°$

17. a. $62° + \left(\frac{55}{60}\right)° \approx 62.92°$

  b. $315° + \left(\frac{20}{60}\right)° \approx 315.33°$

  c. $25° + \left(\frac{25}{60}\right)° \approx 25.42°$

19. a. $128° + \left(\frac{10}{60}\right)° + \left(\frac{40}{3600}\right)° \approx 128.178°$

  b. $13° + \left(\frac{30}{60}\right)° + \left(\frac{40}{3600}\right)° \approx 13.514°$

  c. $48° + \left(\frac{28}{60}\right)° + \left(\frac{10}{3600}\right)° \approx 48.469°$

21. a. (See Example 2 for one step solution.) $16°42'' = 16.012°$

 b. $29°17'' = 29.005°$

 c. $143°23'' = 143.006°$

23. If $\theta = 60°$, then $\beta = 90° - \theta = 90° - 60° = 30°$.

25. If $\phi = 45°$, then $\gamma = 90° - \phi = 90° - 45° = 45°$.

27. $\alpha + \beta + \gamma = 180°$ so $\alpha = -\gamma - \beta + 180°$.

29. $\alpha + \beta + \gamma = 180°$ so $\gamma = -\alpha - \beta + 180°$.

31. If $\theta + \phi = 100°$ and $\theta = \phi$, then $\theta + \theta = 100°$ or $2\theta = 100°$. So $\theta = 50°$ and $\phi = 50°$. Thus, $\beta = 180° - 90° - \theta = 90° - 50° = 40°$.

33. Since $a^2 + b^2 = c^2$, then by substitution $c^2 = 2^2 + 2^2 = 8$. Thus, $c = \sqrt{8} = 2\sqrt{2}$.

35. Since $a^2 + b^2 = c^2$, then by substitution $b^2 = (\sqrt{41})^2 - 5^2 = 16$. Thus, $c = \sqrt{16} = 4$.

37. Since $a^2 + b^2 = c^2$, then by substitution $x^2 + (x+7)^2 = 13^2$. Thus, $2x^2 + 14x - 120 = 0$, and then $x^2 + 7x - 60 = 0$. By factoring, we can see that $(x-5)(x+12) = 0$, so $x = 5$. Thus, we know that $a = 5$, $b = 12$, and $c = 13$.

39. Since $a^2 + b^2 = c^2$, then by substitution $(x-1)^2 + x^2 = (\sqrt{13})^2$ and $x^2 - 2x + 1 + x^2 = 13$ which gives $2x^2 - 2x - 12 = 0$. Since the equation is in polynomial form, we can factor to find the solutions. So

$$2(x^2 - x - 6) = 0 \quad \text{and} \quad 2(x-3)(x+2) = 0.$$

Then the solutions are $x = 3$ or $x = -2$. However, since the length of a side of a triangle cannot be negative, we disregard the negative answer and the correct answer is $x = 3$. The lengths of the legs are $a = 2$ and $b = 3$.

43. We know that $a^2 + b^2 = c^2$ and can use this to determine the length of the roof by solving for c in hte Pythagorean Theorem. Substituting, we get $18^2 + 24^2 = c^2$ which implies that $c^2 = 900$ and then we see that the length of the roof is $c = 30$ feet.

Problem Set 1.2

1. The standard way to label triangles is by labeling vertices A, B, and C where C is the right angle (if one exists), and the opposite sides from the vertices are labeled a, b, and c, where c is the hypotenuse. The measures of angles A, B, and C are labeled α, β, and γ, respectively.

3. C corresponds to F, B corresponds to D, A corresponds to E, CB corresponds to FD, AB corresponds to ED, AC corresponds to EF

5. U corresponds to S, T corresponds to R, R corresponds to T, RS corresponds to TU, ST corresponds to UR, RT corresponds to RT

7. Yes, since two angles are given in both triangles.

9. No, since two corresponding angles are not given in both triangles.

11. Yes, since the angles of each triangle can be found by using the rule $\alpha + \beta + \gamma = 180°$. Then it can be seen that two corresponding angles are given in both triangles.

13. We know that $\dfrac{a}{a'} = \dfrac{b}{b'}$. Then when we substitute, we get $\dfrac{4}{2} = \dfrac{8}{b'}$ and then $4b' = 16$. Consequently, $b' = 4$.

15. We know $\dfrac{a'}{a} = \dfrac{c'}{c}$, and when we substitute, we see that $\dfrac{a'}{4} = \dfrac{8}{6}$ and then $6a' = 32$. Thus, $a' = \dfrac{16}{3}$.

17. We know $\dfrac{b}{b'} = \dfrac{c}{c'}$, and we can substitute to get $\dfrac{b}{8} = \dfrac{4}{12}$. Then we know that $12b = 32$ and thus $b = \dfrac{8}{3}$.

19. Using the Pythagroean Theorem, we can substitute the values into the equation and then solve for a, the distance from the base of the building to the base of the ladder. Doing so yields $a^2 + 10^2 = 26^2$, and then $a^2 = 576$. Thus, the distance is $a = 24$ feet.

21. Using the Pythaogean Theorem, we can sbstute values and solve for the length of the diagonal, c. Thus, $6^2 + 8^2 = c^2$ and $c^2 = 100$. Thus, the diagonal is $c = 10$ feet.

23. By the Pythagorean Theorem, $3^2 + 3^2 = c^2$, where c is the hypotenuse. Thus, $c^2 = 18$ and then $c = \sqrt{18} = 3\sqrt{2}$.

25. Substituting into the Pythagorean Theorem, we see that $10^2 + 15^2 = c^2$, where c is the length of the guy wire. Then, $c^2 = 325$ and $c = 5\sqrt{13}$ feet exactly, and approximately 18 feet. The amount of wire that would need to be bought would be $18 + 18 + 18 + 1 = 55$ feet, since $5\sqrt{13}$ is slightly larger than 18.

27. Using similar triangles, we know that $\dfrac{AB}{DE} = \dfrac{AC}{DC}$. Substituting, we get $\dfrac{AB}{4} = 53$ and then $AB = \dfrac{20}{3}$.

29. We know that $\dfrac{x}{350} = \dfrac{50}{140}$ and then $x = \dfrac{17500}{140}$ and consequently $x = 125$ feet.

31. We can use similar triangles to determine the hieght of the tree. We use $\dfrac{x}{6} = \dfrac{12}{2.5}$, and then $x = \dfrac{72}{2.5} = 28.8$ feet.

Problem Set 1.3

1. If θ is an acute angle in a right triangle, then

$$\cos\theta = \frac{\text{ADJACENT SIDE}}{\text{HYPOTENUSE}}, \quad \sin\theta = \frac{\text{OPPOSITE SIDE}}{\text{HYPOTENUSE}}, \quad \tan\theta = \frac{\text{OPPOSITE SIDE}}{\text{ADJACENT SIDE}}$$

3. The agreement is to carry out all calculations without rounding. After obtaining your final answer, round this answer to be as accurate as the least precise measurement.

5. a. $\cos\alpha = \frac{1}{2}$, $\sin\alpha = \frac{\sqrt{3}}{2}$, and $\tan\alpha = \sqrt{3}$.

    b. $\cos\beta = \frac{\sqrt{3}}{2}$, $\sin\beta = \frac{1}{2}$, and $\tan\alpha = \frac{1}{\sqrt{3}}$.

7. a. 0.9903   b. 0.4067   c. 1.8807

9. a. -0.5774 b. -0.3420 c. 0.7660

11. a. -0.6428 b. -1.5557 c. 0.4132

13. a. 0.6561   b. 0.6561

15. a. 0.9063   b. -0.9063

17. a. $0.9988 + 0.0012 = 1$ b. $0.3290 + 0.6710 = 1$ c. $0.8600 + 0.1403 = 1.003$.

19. 53°

21. 31°

23. 68°

25. 24°

27. a. 60°   b. 1.00381983754

29. a. 50.1944289077°   b. 47.7395014064

31. 0.991386380513

33. 0.500055026568

35. $\alpha = 69°$, $\beta = 21°$, $\gamma = 90°$, $a = 96$, $b = 37$, $c = 100$

37. $\alpha = 77°$, $\beta = 13°$, $\gamma = 90°$, $a = 390$, $b = 90$, and $c = 400$

39. $\alpha = 50°$, $\beta = 40°$, $\gamma = 90°$, $a = 98$, $b = 82$, and $c = 128$

41. $\alpha = 67.8°$, $\beta = 22.2°$, $\gamma = 90°$, $a = 26.6$, $b = 10.8$, and $c = 28.7$

43. $\alpha = 28.95°$, $\beta = 61.05°$, $\gamma = 90°$, $a = 202.7$, $b = 366.4$, and $c = 418.7$

45. $\alpha = 56°$, $\beta = 34°$, $\gamma = 90°$, $a = 3484$, $b = 2350$, and $c = 4202$

47. $\alpha = 57.83°$, $\beta = 32.17°$, $\gamma = 90°$, $a = 290.9$, $b = 182.9$, and $c = 343.6$

49. $\alpha = 28°$, $\beta = 62°$, $\gamma = 90°$, $a = 1600$, $b = 3100$, and $c = 3500$

51. $\alpha = 34.2°$, $\beta = 55.8°$, $\gamma = 90°$, $a = 85.3$, $b = 125.5$, and $c = 151.7$

53. $\alpha = 15.3°$, $\beta = 74.7°$, $\gamma = 90°$, $a = 12.5$, $b = 45.6$, and $c = 47.3$

55. $\alpha = 28°$, $\beta = 62°$, $\gamma = 90°$, $a = 135$, $b = 256$, and $c = 289$

57. $\alpha = 30.5°$, $\beta = 59.5°$, $\gamma = 90°$, $a = 111.7$, $b = 189.4$, and $c = 219.9$

59. $\tan\theta = \dfrac{PA}{DA}$, thus $\tan\theta = \dfrac{PA}{50}$ so $PA = 32$.

61. $\sin 52° = \frac{h}{16}$ so $h = 12.6$. The top of the ladder is 13 ft above the ground.

65. $\tan 51.36° = \frac{h}{1000}$ so $h = 1250.89$ The height of the chimney is 1251 ft.

67. We know that for right traingles, $\tan\theta = \dfrac{OPP}{ADJ}$, so we can substitute the values in to get $\tan\theta = \dfrac{70}{98}$, where we have converted the measurements to inches. Thus, the angle above the horizon the sun is on is $\theta = 35.5° \approx 36°$.

Problem Set 1.4

1. The angle of elevation is the acute angle measured up from a horizontal line to the line of sight.

3. Bearing is the acute angle made with a north-south line. $S38.5°W$ means that we start on the north-south line and move the terminal line 38.5° to the left of the south branch. We cannot write $W38.5°S$ because this implies that we should start on a east-west line and moving 38.5° to the south of the west branch.

5. We can find the height of the building by solving $\tan 38° = \dfrac{x}{150}$ for x. We get $x = 150\tan 38°$, and thus $x = 117.2$ feet.

7. The distance from the car to the helicoptor can be found by solving $\cos 53.4° = \dfrac{1250}{x}$ for x. Thus, we can see $x = \dfrac{1250}{\cos 53.4°}$ and thus $x = 2097$ feet.

9. We can determine the amount of fence needed by solving for the two other sides of the fence and adding the sides. Then we need to solve the equations, $\sin 29.5° = \dfrac{b}{35}$ and $\cos 29.5° = \dfrac{a}{35}$. These give us $b = 17.2$ feet and $a = 30.5$. Then the amount of fence needed is $35 + 17.2 + 30.5 = 82.7 \approx 83$ feet.

17. $\tan 49° = \dfrac{CB}{300}$ and $CB = 345.1$. The distance from C to B is 350 ft.

19. The circumference of the wheel is $2\pi(2.5) = 5\pi$. This becomes the hypotenuse of the triangle we need to solve. Then we need to solve for the height of the triangle, x. Then we examine $\sin 15° = \dfrac{x}{5\pi}$. This yields $x = 5\pi \sin 15° \approx 4.1$ feet. Thus the height of the center of the wheel is $4.1 + 2.5 = 6.6$ feet.

21. Since $\tan 55.48° = \dfrac{h}{1000}$ the height of the Sears Tower is $h \approx 1450$ ft. The angle of depression from the top of the Sears Tower to the top of the Empire State Building is $11.53°$ and $\tan 11.53° = \frac{h}{1000}$ so the height difference is $h \approx 204$ ft. So, the Empire State Building is $1450 - 200 = 1250$ ft.

23. The distance across the river is 173 m.

25. From problem 42, we know that 995 m away from the base of Devil's Tower, the angle of elevation is $14.8°$, so we can see that $\tan 14.8° = \dfrac{h}{995}$ and the height of Devil's Tower is $h \approx 263$ ft.

27. Using the measurements from the diagram, we can solve $\cos 47.0° = \dfrac{x}{92900000}$ for x. We get $x = 92900000 \cos 47.0° = 63,400,000$ mi.

29. To show the statement is true, we can substitute the values of $\tan \alpha = \dfrac{h}{x}$ and $\tan \beta = \dfrac{h}{d+x}$ and see if the statement holds, which it does.

31. Since the difference between the sun's times is 6 h 3 min. Since $30°$ is a third of $90°$, we can multiply the time by a third to get 2 h 1 min. Adding this to the sunrise time, we get that the time would be 8:08 A.M.

33. a. Solving the equation $\tan \theta = \dfrac{6}{4.2}$ for θ gives us that $\theta = 88°$.

   b. Using the same idea as Problem 31, we can see that 5 h 52 min has passed for the sun to reach $88°$ in the sky. Thus, the time would be 12:07 P.M.

35. The two end triangles of the truss have measures of $\gamma = 90°, \alpha = 45°, \beta = 45°, a = 10, b = 10, c = 10\sqrt{2}$. The four interior triangles are congruent with measures $\gamma = 90°, \alpha = 39.8°, \beta = 50.2°, a = 10, b = 12, c = 2\sqrt{61}$.

37. Given that the triangle is drawn as indicated, we can see that $\dfrac{\sin \alpha}{a} = \dfrac{\frac{d}{c}}{a} = \dfrac{d}{ac} = \dfrac{\frac{d}{a}}{c} = \dfrac{\sin \gamma}{c}$.

39. Given that the triangle is drawn as indicated, we can see that $\dfrac{\sin\beta}{b} = \dfrac{\frac{d}{c}}{b} = \dfrac{d}{cb} = \dfrac{\frac{d}{b}}{c} = \dfrac{\sin\gamma}{c}$.

Problem Set 1.5

5. a. quadrant II   b. 45°   c. none   d. −225°

7. a. quadrant III   b. 20°   c. none   d. −160°

9. a. quadrant III   b. 45°   c. 225°   d. none

11. a. quadrant II   b. 20°   c. 160°   d. none

13. a. None, the angle lies on the y-axis.   b. 90°

      c. none   d. −270°

15. a. quadrant IV   b. 30°   c. 330°   d. None

17. a. reference angle

      b. coterminal

      c. coterminal

      d. none

      e. coterminal

23. The equation of the circumference of a circle is $C = 2\pi r$. We substitute $r = 3$ and we have a circumference of 6π in. Using the formula $\dfrac{\theta}{360°} = \dfrac{s}{C}$, we substitute $\theta = 360°$ and $\dfrac{360°}{360°} = \dfrac{s}{6\pi}$ so, $1 = \dfrac{s}{6\pi}$. Then multiply both sides by 6π and $s = 6\pi \approx 18.85$ in.

25. The equation of the circumference of a circle is $C = 2\pi r$. We substitute $r = 10$ and we have a circumference of 20π ft. Using the formula $\dfrac{\theta}{360°} = \dfrac{s}{C}$, we substitute $\theta = 90°$ and $\dfrac{90°}{360°} = \dfrac{s}{6\pi}$ so, $\dfrac{1}{4} = \dfrac{s}{6\pi}$. Then multiply both sides by 6π and $s = \dfrac{3\pi}{2} \approx 4.71$ ft.

27. $\dfrac{112°}{360°} = \dfrac{s}{14.4\pi}$ then $s = (14.4\pi)\left(\dfrac{14}{45}\right) \approx 14.07$ cm.

29. $\dfrac{48°}{360°} = \dfrac{s}{84\pi}$ then $s = (84\pi)\left(\dfrac{2}{15}\right) \approx 35.19$ ft.

31. The hour hand of a clock moves 90° in 3 hours. Using the arc length formula, we substitute $\theta = 90°$ and $\dfrac{90°}{360°} = \dfrac{s}{4\pi}$ so, $\dfrac{1}{4} = \dfrac{s}{4\pi}$. Then multiply both sides by 4π cm and $s = \dfrac{4\pi}{4} = \pi \approx 3.14$ cm.

33. A 20.0 cm diameter wheel has a radius of 10.0 cm. Also, 1.00 meter = 100 cm is the value of the arc length, s. Let the unknown central angle be θ. From $s = r\theta$, we get $\theta = \dfrac{s}{r}$. So, $\theta = \dfrac{100}{10} = 10$. The pulley turns through an angle of 10 radians which is approximately 1.59 rev.

35. $\theta = 118° - 81° = 37°$. Convert degrees to radians, $(37°)(\frac{\pi}{180°}) = 0.646$. Then $s = r\theta = (3950)(0.646) \approx 2600$ mi.

37. $\theta = 33° - 25° = 8°$. Convert degrees to radians, $(8°)(\frac{\pi}{180°}) = 0.13963$. Then
$s = r\theta = (6370)(0.13963) \approx 889$ km. Rounded to the nearest 10 km, the distance is 890 km.

39. $\theta = 45.57' = 0.7625°$ and $r = 384417$ km $+ 6370$ km $= 390787$ km. Then
$s = r\theta = (390787)(0.7625°)(\frac{\pi}{180°}) \approx 5200$ km is the diameter of the moon.

Chapter 1 Sample Test

1. a. alpha   b. theta   c. gamma   d. lambda   e. beta   f. delta

3. $50°36' = 50.60°$.

5. a. 0.9397   b. 0.6428   c. −0.2679

7. We can find the height of the tower by solving the equation $\tan 76.35° = \dfrac{x}{501.0}$ for x. This
gives us $x = 501.0 \tan 76.35° = 2036$ ft.

Chapter 1 Miscellaneous Problems

1. Use $s = r\theta$ and $s = (5$ in.$)(6) = 30$ in.

3. Use $s = r\theta$ and $s = (6.4$ ft$)(1.13) = 7$ ft.

5. $375° - 360° = 15°$ so $375°$ is a quadrant I angle coterminal with $15°$.

7. $815° - 2(360°) = 95°$ so $815°$ is a quadrant II angle coterminal with $95°$.

9. $\sin 85.2° = 0.9965$

11. 0.9994

13. 0.8192

15. 1.7321

17. $-\tan 34.5° = -0.6873$

19. 0.7720

21. $a = 1.8, b = 3.6, c = 4.0, \alpha = 27°, \beta = 63°, \gamma = 90○$

23. $a = 7.5, b = 22.0, c = 23.2, \alpha = 71.2°, \beta = 18.8°, \gamma = 90○$

25. $a = 2.71, b = 6.5, c = 7.0, \alpha = 21.8°, \beta = 68.2°, \gamma = 90\circ$

27. $a = 5.81, b = 3.91, c = 7.00, \alpha = 56.0°, \beta = 34.0°, \gamma = 90\circ$

29. $a = 2.2, b = 8.2, c = 8.5, \alpha = 14°, \beta = 76°, \gamma = 90\circ$

31. $a = 7.8, b = 20.12, c = 50.05, \alpha = 66.3°, \beta = 23.7°, \gamma = 90\circ$

33. Use $s = r\theta$ where $r = 3950$ mi and $\theta = (1°)\left(\frac{\pi}{180°}\right)$. Thus, $s = (3950$ mi $)\left(\frac{1}{60} \cdot \frac{\pi}{180}\right) = 1.149008424$ mi.

35. After a few repititions of the process of solving for c in the Pythagorean Theorem, one can see the pattern forming that the length of the hypotenuse of the nth triangle can be represented by $\sqrt{n+1}$, so in the case of the sixth triangle, the length is $\sqrt{7}$.

37. 42°N latitude must first be converted to radians. Once converted, this will be the central angle θ needed to compute the arc length. Since latitude is the angular measure of the point north or south of the equator, latitude determines the angle needed to compute the distance to the equator. Longitude is not needed. $\theta = (42°)(\frac{\pi}{180°}) = 0.733$ and $s = r\theta = (3950$ mi$)(0.733) = 2895$ mi. Thus, it is about 2900 miles from Chicago to the equator.

39. First we compute the circumference of the Earth given the radius, that is, $C = 2r\pi = 24818.6$ mi. We then compute the fraction of the surface that the distance from Sacramento and Santa Rosa composes, that is, $\frac{90}{23818.6} = 0.003626315159$. Using this ratio, we can find the fraction of time (of 24 h) it takes for the line of sight of the sun to pass between the cities, that is, $(0.003626315159)(24) = 0.087031563817$ h. This computes to approximately five min.

41. We can find the height of the sign by finding the height from the top of the sign and the height from the bottom of the sign, and subtracting the two. Thus, we need to compute $H_{top} - H_{bottom} = (275\tan 21.5°) - (275\tan 18.5°) = 108.325 - 92.0137 \approx 16.3$ ft.

Problem Set 2.1

1. $30° = \frac{\pi}{6}$ not $\frac{\pi}{3}$.

3. These angles are not equal because a positive number cannot equal a negative number.

5. The circumference of a circle is found by the formula $C = 2\pi r$. If a circle has a radius of 10, then the circumference is $C = (2\pi)(10) = 20\pi$ not 10π.

7. a. $(30°)(\frac{\pi}{180°}) = \frac{\pi}{6}$

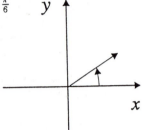

b. $(90°)(\frac{\pi}{180°}) = \frac{\pi}{2}$

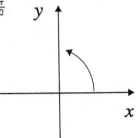

c. $(45°)(\frac{\pi}{180°}) = \frac{\pi}{4}$

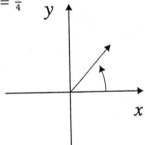

d. $(60°)(\frac{\pi}{180°}) = \frac{\pi}{3}$

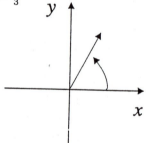

e. $(180°)(\frac{\pi}{180°}) = \pi$

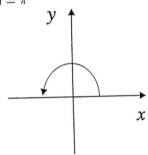

f. $(270°)(\frac{\pi}{180°}) = \frac{3\pi}{2}$

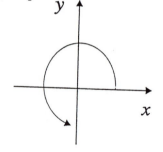

9.   a. $\left(\frac{\pi}{3}\right)\left(\frac{180°}{\pi}\right) = 60°$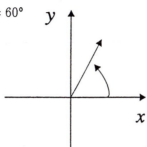

b. $\left(\frac{\pi}{6}\right)\left(\frac{180°}{\pi}\right) = 30°$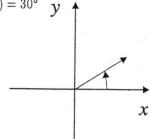

c. $\left(\frac{\pi}{2}\right)\left(\frac{180°}{\pi}\right) = 90°$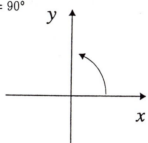

d. $\left(\frac{\pi}{4}\right)\left(\frac{180°}{\pi}\right) = 45°$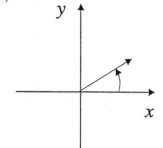

e. $(2\pi)\left(\frac{180°}{\pi}\right) = 360°$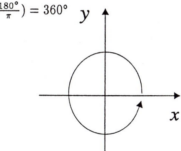

f. $(\pi)\left(\frac{180°}{\pi}\right) = 180°$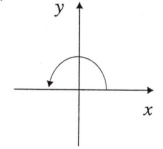

11.   a. $\frac{\pi}{3} \approx 1.04$ so the answer is F.

b. $\frac{\pi}{12} \approx 0.26$ so the answer is D.

c. $\frac{\pi}{180} \approx 0.01$ so the answer is A.

d. $\frac{\pi}{4} - \frac{1}{4} \approx 0.53$ so the answer is E.

e. $\frac{1}{3}\frac{\pi}{10} \approx 0.10$ so the answer is C.

f. $\frac{1}{50}\pi \approx 0.06$ so the answer is B.

13.   a.   reference angle

b.   none apply

c.   coterminal

    d. coterminal

    e. equal

    f. coterminal

15. a. reference angle

    b. equal

    c. none apply

    d. none apply

    e. none apply

    f. coterminal

17. a. $(150°)(\frac{\pi}{180°}) = \frac{5\pi}{6}$

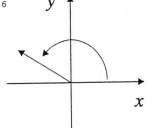

    b. $(135°)(\frac{\pi}{180°}) = \frac{3\pi}{4}$

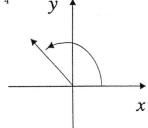

    c. $(20°)(\frac{\pi}{180°}) = \frac{\pi}{9}$

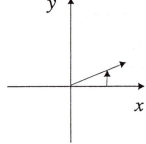

19.   a. $(225°)(\frac{\pi}{180°}) = \frac{5\pi}{4}$

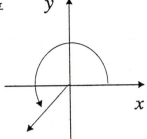

b. $(-240°)(\frac{\pi}{180°}) = -\frac{4\pi}{3}$

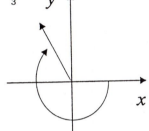

c. $(250°)(\frac{\pi}{180°}) = \frac{25\pi}{18}$

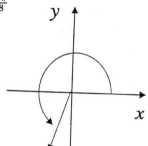

21.   a. $(-60°)(\frac{\pi}{180°}) \approx -1.05$

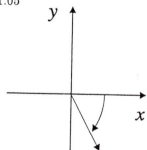

b. $(400°)\left(\frac{\pi}{180°}\right) \approx 6.98$

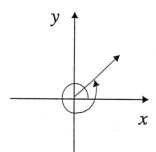

c. $(23.7°)\left(\frac{\pi}{180°}\right) \approx 0.41$

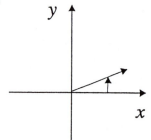

23.   a. $(38.4°)\left(\frac{\pi}{180°}\right) \approx 0.67$

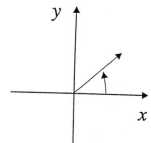

b. $(-210°)\left(\frac{\pi}{180°}\right) \approx -3.67$

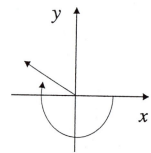

c. $(-825°)(\frac{\pi}{180°}) \approx -14.40$

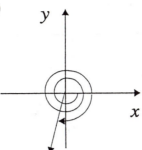

25.   a. $(-\frac{5\pi}{3})(\frac{180°}{\pi}) = -300°$

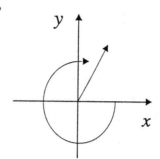

b. $(\frac{5\pi}{3})(\frac{180°}{\pi}) = 300°$

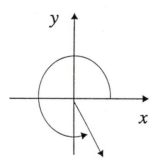

c. $(-2\pi)(\frac{180°}{\pi}) = -360°$

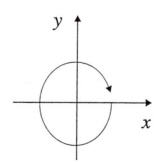

27. a. $\left(-\frac{\pi}{4}\right)\left(\frac{180°}{\pi}\right) = -45°$

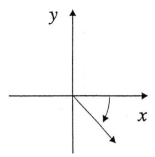

  b. $\left(\frac{5\pi}{4}\right)\left(\frac{180°}{\pi}\right) = 225°$

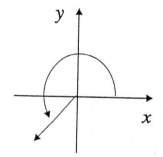

  c. $\left(-\frac{11\pi}{4}\right)\left(\frac{180°}{\pi}\right) = -495°$

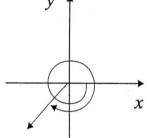

29. a. $2\left(\frac{180°}{\pi}\right) = \left(\frac{360°}{\pi}\right) \approx 115°$

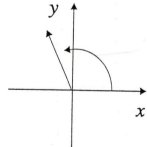

b. $-3\left(\frac{180°}{\pi}\right) = \frac{-540°}{\pi} \approx -172°$

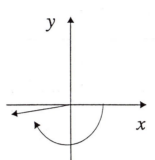

c. $0.5\left(\frac{180°}{\pi}\right) = \frac{90°}{\pi} \approx 29°$

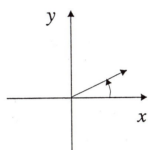

31.   a. $12\left(\frac{180°}{\pi}\right) = \left(\frac{2160°}{\pi}\right) \approx 327°$

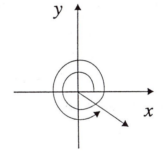

b. $-1.5\left(\frac{180°}{\pi}\right) = \frac{-270°}{\pi} \approx -86°$

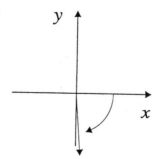

c. $(4.712389)(\frac{180°}{\pi}) = \frac{848.23002°}{\pi} \approx 270°$

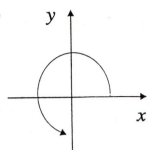

33. a. $\dfrac{7\pi}{4}$ b. $\dfrac{\pi}{4}$ c. $\dfrac{5\pi}{3}$

d. $2\pi - 2$ e. $8 - 2\pi$ f. $\dfrac{11\pi}{6}$

35. a. $-6 + 2\pi \approx 0.2832$

b. $6.2832 - 2\pi \approx 0.0000$

c. $3\sqrt{5} - 2\pi \approx 0.4250$

37. a. $-0.7854 + 2\pi \approx 5.4978$

b. $6.8068 - 2\pi \approx 0.5236$

c. $\frac{9\pi}{4} - 2\pi \approx 0.7854$

39. a.   Since $s = r\theta$, then $s = (6 \text{ cm })(2.34) = 14.04$ cm.

b.   Since $s = r\theta$,then $s = (15 \text{ cm })(\frac{2\pi}{3}) = 31.42$ cm.

41. Since $\omega = \frac{\theta}{t}$ and $v = r\omega$, then $\omega = \frac{18\pi}{5}$ and $v = (5)(\frac{18\pi}{5}) = 18\pi$ ft/min.

43. 365 days are equivalent to 8760 hours. Thus, $\omega = \frac{1}{8760 \text{h}}$ and $v = (93000000 \text{ mi })(\frac{1}{8760 \text{ h}}) \approx$ 10600 mi/h.

45. Since $s = r\theta$, then $s = (10 \text{ in })(4.5\pi) \approx 141.4$ in.

47. Since the wheel has a diameter of 60 cm, it has a radius of 30 cm. Recall that $\omega = \frac{v}{r}$, and here the linear velocity given is 30 km/h. Thus, $\omega = \frac{30}{0.0003} = 100000$ per hour. Now, convert to rad/s, so $\omega = (100000)(\frac{1}{3600}) \approx 28$ rad/s.

49. $s = r\theta = (2 \text{ cm})(2) = 4$ cm.

51. Since the bike wheel is making 3 revolutions per second, it's angular velocity is $\omega = (3$ rev/s$)(2\pi \text{ rad}) = 6\pi$ rad/s. The linear velocity is given by the equation $v = \omega r = (6\pi)(\frac{11}{12}) \approx$ 17 ft/s.

53. The wheel spocket times the number of rotations is equal to the drive spocket times the number of rotations. So, $(4)(x) = (10)(1)$, and $x = 2.5$ rotations of the wheel for one rotation of the drive spocket.

55. $v = 30$ mi/h $= 528$ in/s. The angular velocity is given by the equation $\omega = \frac{v}{r} = \frac{528}{15} = 35.2$ rad/s.

57. The angular velocity is given in rev/min and must be converted to rad/hr, so $\omega = 40$ rev/min $= 4800\pi$ rad/hr and $v = r\omega = (\frac{125}{5280})(4800\pi) \approx 357$ mi/hr.

59. Once again use a ratio comparison between the two spockets, so $\dfrac{r_1\theta_1}{r_2} = \theta_2$.

Problem Set 2.2

1. Let θ be an angle in standard position with the point (a, b) the intersection of the terminal side of θ and the unit circle. The six trigonometric functions are defined as follows: $\cos\theta = a$; $\sin\theta = b$; $\tan\theta = \dfrac{b}{a}, a \neq 0$; $\cot\theta = \dfrac{a}{b}, b \neq 0$; $\sec\theta = \dfrac{1}{a}, a \neq 0$; $\csc\theta = \dfrac{1}{b}, b \neq 0$.

3. $(\cos\beta, \sin\beta)$

5. $[\cos(\beta + \pi), \sin(\beta + \pi)]$

7. The function has been evaluated in radians instead of degrees, the correct answer is 0.5.

9. $\cos^{-1}\theta$ is a degree, and $\dfrac{1}{\csc\theta}$ is a number.

11. The sine is negative in quadrant VI, not positive.

13. The sine is positive and the cosine is negative in quadrant II.

15. a. $\cos 50° = 0.6$ b. $\sin 70° = 0.9$ c. $\sin -150° = -0.5$

17. a. $\cos 2 = -0.4$ b. $\sin 4 = -0.8$ c. $\tan -6 = 0.3$

19. a. $\cos 50° = 0.64$ b. $\sin 20° = 0.34$

   c. $\sec 70° = 2.92$ d. $\cos -34° = 0.83$

   e. $\sin -95° = -0.10$ f. $\cot 250° = 0.36$

21. a. $\cos\left(\dfrac{103°}{2}\right) = 0.62$ b. $\dfrac{\cos 103°}{2} = -0.11$

   c. $\dfrac{\cot 103°}{2} = -0.12$ d. $\cos\dfrac{-5\pi}{4} = \dfrac{-\sqrt{2}}{2}$

   e. $\sin\dfrac{-5\pi}{4} = \dfrac{\sqrt{2}}{2}$ f. $-\cot\dfrac{-5\pi}{4} = 1$

23. a. $\dfrac{1}{\csc 2} = 0.909297426826$

   b. $\dfrac{1}{\sec 3.5} = -0.936456687291$

    c. $\frac{1}{\cot 4.5} = 4.63733205455$

    d. $1 - \cos^2(28°) = 0.220403548265$

    e. $-\sqrt{1 - \cos^2(190°)} = -0.173648177667$

25. a. $\frac{\sin 50°}{\cos 50°} = 1.19175359259$ b. $\frac{\cos 5}{\sin 5} = -0.295812915533$ c. $\frac{\cos 200°}{\sin 200°} = 2.7474771945$

    d. $(\cos 214°)(\sec 128°) = 1.34658023245$ e. $\sin \frac{\pi}{9} \csc \frac{\pi}{9} = 1$

27. a. positive   b. positive   c. negative   d. positive

29. a. negative   b. negative   c. negative

31. a. negative   b. negative   c. negative

33. a. quadrants I, IV   b. quadrants I, II

35. a. quadrants II, III   b. quadrants II, IV

37. a. quadrant II   b. quadrant III

39. $\sec \theta = \frac{1}{a} = \frac{1}{\cos \theta}$

41.

θ	1	0.5	0.1	0.01	0.001	0.0001
$(\sin \theta)/\theta$	0.84147	0.95885	0.998334	0.99998	0.99999	0.99999

Problem Set 2.3

1. $\sec \theta = \frac{1}{\cos \theta}, \csc \theta = \frac{1}{\sin \theta}, \cot \theta = \frac{1}{\tan \theta}$

3. $\cos^2 \theta + \sin^2 \theta = 1, \tan^2 \theta + 1 = \sec^2 \theta, 1 + \cot^2 \theta = \csc^2 \theta$

5. $\sin^2 \theta = 1 - \cos^2 \theta, \sin \theta = \pm\sqrt{1 - \cos^2 \theta}$

7. $\cot^2 \theta = \csc^2 \theta - 1, \cot \theta = \pm\sqrt{\csc^2 \theta - 1}$

9. $\cos = \frac{a}{u}, \sec = \frac{u}{a}, \sin = \pm\sqrt{1 - \frac{a}{u}^2} \csc = \frac{1}{\sqrt{1 - \frac{a}{u}^2}} \tan = \frac{u\sqrt{1 - \frac{a}{u}^2}}{a} \cot = \frac{a}{u\sqrt{1 - \frac{a}{u}^2}}$

11. 0.990

13. 1.94

15. 0.26

17. −0.78

19. 0.77

21. -0.17

23. -1

25. $\frac{1}{2}$

27. $-\frac{1}{2}$

29. $\cos\theta = \dfrac{\pm 1}{\sqrt{1+\tan^2\theta}}$;

$\sin\theta = \dfrac{\pm\tan\theta}{\sqrt{1+\tan^2\theta}}$;

$\tan\theta = \tan\theta$;

$\sec\theta = \pm\sqrt{1+\tan^2\theta}$;

$\csc\theta = \dfrac{\pm\sqrt{1+\tan^2\theta}}{\tan\theta}$;

$\cot\theta = \dfrac{1}{\tan\theta}$

31. $\cos\theta = \dfrac{1}{\sec\theta}$;

$\sin\theta = \dfrac{\pm\sqrt{\sec^2\theta-1}}{\sec\theta}$;

$\tan\theta = \pm\sqrt{\sec^2\theta-1}$;

$\sec\theta = \sec\theta$;

$\csc\theta = \dfrac{\pm\sec\theta}{\sqrt{\sec^2\theta-1}}$;

$\cot\theta = \pm\sqrt{\csc^2\theta-1}$

33. Since $\cos\theta = \frac{5}{13}$ and $\sin^2\theta + \cos^2\theta = 1$, $\sin\theta = \pm\sqrt{1 - \left(\frac{5}{13}\right)^2}$.

35. Since $\sin\theta = 0.65$ and $\sin^2\theta + \cos^2\theta = 1$, $\cos\theta = \pm\sqrt{1 - (0.65)^2}$.

37. Let $u = a\tan\theta$, thus $\dfrac{a\sec^2\theta}{\sqrt{u^2 + a^2}} = \dfrac{a\sec^2\theta}{\sqrt{a^2\tan^2\theta + a^2}} = \dfrac{a\sec^2\theta}{a\sqrt{\tan^2\theta + 1}} = \dfrac{\sec^2\theta}{\sec\theta} = \sec\theta$

39. Let $u = \sqrt{5}\tan\theta$, thus $\dfrac{(\sqrt{5}\tan\theta)(\sqrt{5}\sec^2\theta)}{\sqrt{(\sqrt{5}\tan\theta)^2 + 5}} = \dfrac{5\tan\theta\sec^2\theta}{\sqrt{5}\sqrt{\tan^2\theta + 1}} = \dfrac{5\tan\theta\sec^2\theta}{\sqrt{5}\sec\theta} = \sqrt{5}\tan\theta\sec\theta$

41.

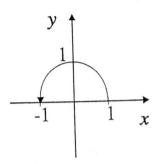

43.

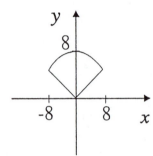

45.

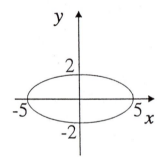

47.

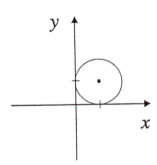

51. Given $v_0 = 300$ ft/s , $\theta = 42°$, $h_0 = 0$, $x = (300\cos 42°)t$ and $y = (300\sin 42°)t - 16t^2$. When $y = 0$ the time of impact is $t = 12.5$ sec.

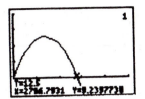

53. $\cot\theta = \dfrac{1}{\tan\theta} = \dfrac{1}{\frac{b}{a}} = \dfrac{a}{b} = \dfrac{\cos\theta}{\sin\theta}$

55. $\left(\frac{a}{b}\right)^2 + 1 = \frac{a^2}{b^2} + \frac{b^2}{b^2} = \dfrac{a^2 + b^2}{b^2} = \dfrac{\cos^2\theta + \sin^2\theta}{\sin^2\theta} = \dfrac{1}{\sin^2\theta} = \csc^2\theta$

Problem Set 2.4

1. Let θ be any angle in standard position, and let $P(x, y)$ be any point on the terminal side of the angle at a distance of r from the origin ($r \neq 0$). Then
$$\cos\theta = \frac{x}{r}; \quad \sin\theta = \frac{y}{r}; \quad \tan\theta = \frac{y}{x} \ (x \neq 0); \quad \sec\theta = \frac{r}{x} \ (x \neq 0); \quad \csc\theta = \frac{r}{y} \ (y \neq 0); \quad \cot\theta = \frac{x}{y} \ (y \neq 0).$$

3. a. $\tan\frac{\pi}{4} = 1$ b. $\cos 0 = 1$ c. $\sin 60° = \frac{\sqrt{3}}{2}$

   d. $\cos 30° = \frac{\sqrt{3}}{2}$ e. $\cos 270° = 0$ f. $\tan\frac{\pi}{6} = \frac{\sqrt{3}}{3}$

5. a. $\cos 60° = \frac{1}{2}$ b. $\sin\frac{\pi}{6} = \frac{1}{2}$ c. $\sin 0° = 0$

   d. $\cos\pi = -1$ e. $\tan\frac{\pi}{3} = \sqrt{3}$ f. $\cos\frac{\pi}{2} = 0$

7. a. $\sec\frac{\pi}{3} = 2$ b. $\cot\frac{\pi}{6} = \sqrt{3}$ c. $\cot 90° = 0$

   d. $\csc 60° = \frac{2}{\sqrt{3}}$ or $\frac{2\sqrt{3}}{3}$ e. $\cot\frac{\pi}{3} = \frac{1}{\sqrt{3}}$ or $\frac{\sqrt{3}}{3}$ f. $\cot 0$ is undefined.

9. a. $\tan 90°$ is undefined.   b. $\sin\frac{3\pi}{2} = -1$ c. $\tan\frac{3\pi}{2}$ is undefined.

   d. $\cot\frac{\pi}{4} = 1$ e. $\csc\frac{\pi}{4} = \sqrt{2}$ f. $\sec 180° = -1$

11. a. $\cos\frac{9\pi}{2} = \cos\frac{\pi}{2} = 0$ b. $\cos 495° = -\cos 45° = -\frac{\sqrt{2}}{2}$

   c. $\sin -765° = -\sin 45° = -\frac{\sqrt{2}}{2}$ d. $\cos 300° = \cos 60° = \frac{1}{2}$

   e. $\sin 120° = \sin 60° = \frac{\sqrt{3}}{2}$ f. $\tan 120° = -\tan 60° = -\sqrt{3}$

13. a. $\sin -390° = -\sin 30° = -\frac{1}{2}$ b. $\tan\frac{11\pi}{4} = -\tan\frac{\pi}{4} = -1$

   c. $\cos -\frac{17\pi}{6} = -\cos\frac{\pi}{6} = -\frac{\sqrt{3}}{2}$ d. $\sec -420° = \sec 60° = 2$

15. $(3, -4)$ To find the radius, $r^2 = x^2 + y^2 = 3^2 + (-4)^2 = 25$, Thus $r = 5$. Thus, $\cos\theta = \frac{3}{5}$; $\sin\theta = -\frac{4}{5}$; $\tan\theta = -\frac{4}{3}$; $\sec\theta = \frac{5}{3}$; $\csc\theta = -\frac{5}{4}$; $\cot\theta = -\frac{3}{4}$.

17. $(-5, 12)$ $r^2 = (-5)^2 + (12)^2 = 169$, Thus $r = 13$. Thus,
$\cos \theta = -\frac{5}{13}$; $\sin \theta = \frac{12}{13}$; $\tan \theta = -\frac{12}{5}$; $\sec \theta = -\frac{13}{5}$; $\csc \theta = \frac{13}{12}$; $\cot \theta = -\frac{5}{12}$.

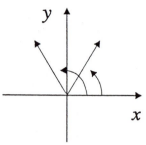

19. $(-2, -3)$ $r^2 = (-2)^2 + (-3)^2 = 13$, Thus $r = \sqrt{13}$. Thus,
$\cos \theta = -\frac{2}{\sqrt{13}}$; $\sin \theta = -\frac{3}{\sqrt{13}}$; $\tan \theta = \frac{3}{2}$; $\sec \theta = -\frac{\sqrt{13}}{2}$;
$\csc \theta = -\frac{\sqrt{13}}{3}$; $\cot \theta = \frac{2}{3}$.

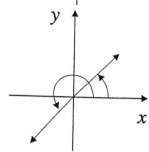

21. False, take $\theta = \frac{\pi}{2}$. The cosine is 0 and the secant is undefined, and they cannot be multiplied.

23. False, take $\theta = 30°$, then $(\sin 30° + \cos 30°)^2 \neq 1$.

25. The statement $\cos \theta = \frac{\sin \theta}{\tan \theta}$ is equivalent to $\tan \theta = \frac{\sin \theta}{\cos \theta}$.

27. False, take $\theta = 30°$, then $\sin -30° \neq \sin 30°$.

29. False, take $\theta = \frac{\pi}{2}$, then $\dfrac{\sin 5 \left(\frac{\pi}{2} \right)}{5} \neq \sin \frac{\pi}{2}$.

31. False, take $\theta = \frac{\pi}{2}$, then $\sin \left(\frac{1}{2} \right) \left(\frac{\pi}{2} \right) \neq \frac{1}{2} \sin \frac{\pi}{2}$.

33. False, take $\theta = \frac{\pi}{2}$, then $\sin \left(\frac{1}{2} \right) \left(\frac{\pi}{2} \right) \neq -\sqrt{\dfrac{1 - \cos \frac{\pi}{2}}{2}}$.

35. False, $\cos \theta = \pm\sqrt{1 - \sin^2 \theta}$.

37. Since $\cos \dfrac{5\pi}{4} = \dfrac{x}{r} = \dfrac{x}{\sqrt{x^2 + y^2}}$. Then $\theta = \dfrac{5\pi}{4}$, so $y = x$ bisects Quadrant III and is negative. So

$$\cos \frac{5\pi}{4} = -\frac{x}{\sqrt{x^2 + x^2}} = -\frac{x}{\sqrt{2x^2}} = -\frac{1}{\sqrt{2}} \quad \text{or} \quad -\frac{1}{2}\sqrt{2}$$

39. Since $\sin\left(-\dfrac{\pi}{4}\right) = \dfrac{y}{r} = \dfrac{y}{\sqrt{x^2 + y^2}}$. Then $\theta = -\dfrac{\pi}{4}$, so $y = -x$ and $(x, -x)$ will be a point on the terminal side. So

$$\sin\left(-\frac{\pi}{4}\right) = -\frac{x}{\sqrt{x^2 + x^2}} = -\frac{x}{\sqrt{2x^2}} = -\frac{1}{\sqrt{2}} \quad \text{or} \quad -\frac{1}{2}\sqrt{2}$$

41. Using the standard position angle $210°$ and also the standard position angle $150°$, we choose $P_1(-x, y)$ and $P_2(-x, -y)$. From geometry, we know that an equiangular triangle has all the sides of the same length. Thus, $2y = r$. So, $r^2 = x^2 + y^2$ and then by substitution, $(2y)^2 = x^2 + y^2$. Hence, $3y^2 = x^2$ or $x = -\sqrt{3}y$ and

$$\cos 210° = \frac{x}{r} = -\frac{\sqrt{3}y}{2y} = -\frac{\sqrt{3}}{2}$$

[Hint: The cosine function is negative in Quadrant III. (See Example 3.)]

43.   a. $2\cos\frac{\pi}{2} = 2(0) = 0$ b. $\cos\frac{2\pi}{2} = \cos\pi = -1$

45.   a. $\sin^2 60° = \frac{3}{4}$ b. $\cos^2\frac{\pi}{4} = \frac{1}{2}$

47.   a. $\sin^2\frac{\pi}{3} + \cos^2\frac{\pi}{3} = 1$ b. $\sin^2\frac{\pi}{6} + \cos^2\frac{\pi}{3} = \frac{1}{4} + \frac{1}{4} = \frac{1}{2}$

49.   a. $\cos(\frac{\pi}{4} - \frac{\pi}{2}) = \cos -\frac{\pi}{4} = \frac{\sqrt{2}}{2}$ b. $\cos\frac{\pi}{4} - \cos\frac{\pi}{2} = \frac{\sqrt{2}}{2}$

51.   a. $\csc(\frac{1}{2} \cdot 60°) = \csc 30° = \dfrac{1}{\sin 30°} = 2$ b. $\dfrac{\csc 60°}{2} = \dfrac{\sqrt{3}}{3}$

53.   a. $\tan(2 \cdot 60°) = \tan 120° = -\sqrt{3}$ b. $\dfrac{2\tan 60°}{1 - \tan^2 60°} = \frac{2\sqrt{3}}{-2} = -\sqrt{3}$

55. The approximate value (for both parts) is 0.0523359562.

57. Since $m = \tan\phi$, then $\phi = 60°$. Thus $m = \tan 60° = \sqrt{3}$.

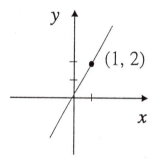

59.   a. $m = \tan 30° = \frac{\sqrt{3}}{3}$ and $y = \frac{\sqrt{3}}{3}(x - 2) + 3$.

   b. $m = \tan 120° = -\sqrt{3}$ and $y = -\sqrt{3}(x - 1) + 4$.

   c. $m = \tan 45° = 1$ and $y = (x - 9) - 5$.

   d. $m = \tan 135° = -1$ and $y = -(x + 3) - 8$.

61.   a. ϕ is between $0 < \phi < \frac{\pi}{2}$ for a line that is rising.

   b. ϕ is between $\frac{\pi}{2} < \phi < \pi$ for a line that is falling.

   c. $\phi = 0°$ or π for all horizontal lines.

   d. $\phi = 90°$ or $\frac{\pi}{2}$ for all vertical lines.

63. $t = 0.06s = \frac{0.06}{60}$ min $= 0.001$ min, $k = 15$, $r = 5$, $v = 600$, and

$$d \approx 15 + 5 - 5\cos[2\pi(600)(0.001)] - \sqrt{15^2 - 5^2 \sin^2[2\pi(600)(0.001)]} \approx 9.3358$$

65. $\cos\theta = \frac{x}{r}$, $\sin\theta = \frac{y}{r}$, so $x = r\cos\theta$ and $y = r\sin\theta$. Then, $P(r\cos\theta, r\sin\theta)$.

Chapter 2 Sample Test

1. $\tan\theta = \frac{\sin\theta}{\cos\theta}$ $\cot\theta = \frac{\cos\theta}{\sin\theta}$

3. $(3.8)(\frac{180°}{\pi}) \approx -218°$

5.   a. $\cos 20° = 0.9397$ b. $\sin 4 = -0.7568$ c. $\tan 0.85 = 1.1383$

      d. $\csc -1.5 = -1.0025$ e. $\sec -85° = 11.4737$

7.   a. $\cos\frac{5\pi}{3} = \cos\frac{\pi}{3}$

      b. $\sin(-\frac{3\pi}{4} = -\sin\frac{\pi}{4}$

      c. $\tan(-\frac{-7\pi}{4} = \tan\frac{\pi}{4}$

      d. $\cot\frac{5\pi}{4} = \cot\frac{\pi}{4}$

      e. $\csc 2 = -\csc(\pi - 2)$

      f. $\sec 4 = -\sec(4 - \pi)$

9.   a. $180°$ b. $-240°$ c. $-\frac{\pi}{4}$ d. $\frac{7\pi}{6}$

      e. $\frac{\pi}{2}$ f. $-\frac{\sqrt{2}}{2}$ g. 0 h. $\frac{\sqrt{3}}{2}$

      i. $-\frac{1}{2}$ j. 1 k. $\frac{\sqrt{2}}{2}$ l. -1

      m. $-\frac{1}{2}$ n. $-\frac{\sqrt{3}}{2}$ o. 0 p. -1

      q. 0 r. $-\sqrt{3}$ s. $\frac{\sqrt{3}}{3}$ t. undefined

Chapter 2 Miscellaneous Problems

1. If $\cos\alpha = \frac{x}{r}$, $\sin\alpha = \frac{y}{r}$, and $r = 2$, then $x = 2\cos\alpha$ and $y = 2\sin\alpha$. So, $(x, y) = (2\cos\alpha, 2\sin\alpha)$.

3. $(150°)(\frac{\pi}{180°}) = \frac{5\pi}{6} \approx 2.62$

5. $(-100°)(\frac{\pi}{180°}) = -\frac{5\pi}{9} \approx -1.75$

7. $\left(\frac{2\pi}{3}\right)\left(\frac{180°}{\pi}\right) = 2(60°) = 120°$

9. $(-2.5)\left(\frac{180°}{\pi}\right) \approx -143°$

11. Use $s = r\theta$ and $s = (5 \text{ in.})(6) = 30$ in.

13. Use $s = r\theta$ and $s = (6.4 \text{ ft})(1.13) = 7$ ft.

15. quadrant II; coterminal angle is $9 - 2\pi$

17. quadrant III; coterminal angle is $-\frac{2\pi}{3} + 2\pi = \frac{4\pi}{3}$

19. $\sin 8.3 = 0.9022$

21. $\cos(-0.42) = 0.9130$

23. $-\sin 6.2 = 0.0831$

25. $\cot 6 = -3.4364$

27. $\csc \frac{\pi}{7} = 2.3048$

29. Using $r = \sqrt{x^2 + y^2} = \sqrt{4^2 + 5^2} = \sqrt{41}$, then $\cos\theta = \frac{4}{\sqrt{41}}$; $\sin\theta = \frac{5}{\sqrt{41}}$; $\tan\theta = \frac{5}{4}$; $\sec\theta = \frac{\sqrt{41}}{4}$; $\csc\theta = \frac{\sqrt{41}}{5}$; $\cot\theta = \frac{4}{5}$.

31. Using $r = \sqrt{x^2 + y^2} = \sqrt{5^2 + (\sqrt{11})^2} = \sqrt{36} = 6$, then $\cos\theta = \frac{5}{6}$; $\sin\theta = \frac{\sqrt{11}}{6}$; $\tan\theta = \frac{\sqrt{11}}{5}$; $\sec\theta = \frac{6}{5}$; $\csc\theta = \frac{6}{\sqrt{11}}$; $\cot\theta = \frac{5}{\sqrt{11}}$.

33. $\frac{\sqrt{3}}{2}$

35. $-\frac{1}{2}$

37. $-\frac{2\sqrt{3}}{3}$

39. $\sqrt{3}$

41. $-\sqrt{3}$

43. undefined

45. $-\sqrt{3}$

47. $-\sqrt{2}$

49.   a. $\omega = 100$ rev/min $= 200\pi$ rad/min.

   b. $v = r\omega = (9)(200\pi)$ in/min $= 5.35$ mi/hr.

   c.   The angular speed of the smaller wheel is about 300 rev/min.

CHAPTER 3   GRAPHS OF TRIGONOMETRIC FUNCTIONS

Problem Set 3.1

1. Both the sine and cosine curves have an amplitude of 1 and a period of 2π. The tangent curve has no amplitude and a period of π.

3.

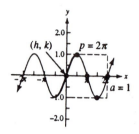

5.

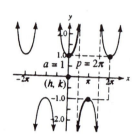

7.

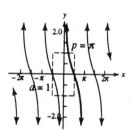

9.

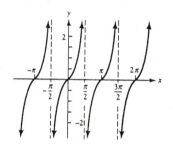

11. Amplitude of 6. Period of 4.

13. Amplitude of 6. Period of 2.

15. Amplitude of 2. Period of $\frac{1}{3}$. Frequency of 3.

17. Amplitude of 0.3. Period of $\frac{1}{6}$. Frequency of 6.

19. The calculator is in degree mode, rather than in radian mode.

21. The calculator is in degree mode, rather than in radian mode.

23.

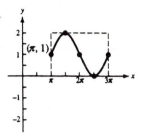

25.

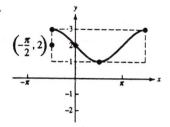

27.

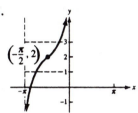

29.

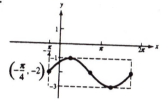

31.

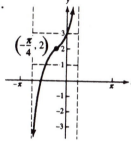

33.

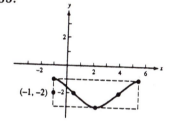

35.

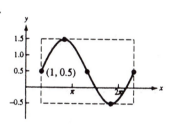

37. $y = \sin x$ has no vertical asymptotes.

39. $y = \sec x$ has vertical asymptotes at $x = \frac{\pi}{2} + n\pi$ for any integer n.

41. $y = \cot x$ has vertical asymptotes at $x = n\pi$ for any integer n.

43. $(-1, -1.19)$, not defined at $x = 0$, $(1, 1.19)$, $(2, 1.10)$, $(3, 7.09)$, $(4, -1.32)$, $(5, -1.04$, $(6, -3.58)$, $(7, 1.52)$

45. a.

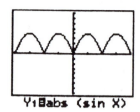

b.

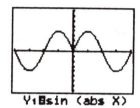

47. a.

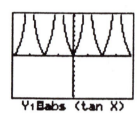

b.

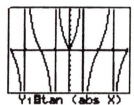

49.

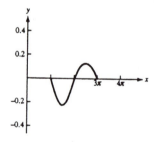

51.

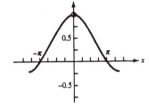

53.
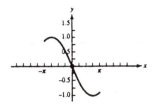

55.
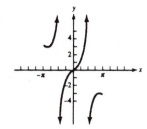

59.
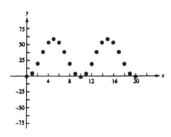

63.
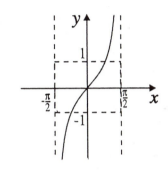

Problem Set 3.2

1. Standard form: $y = \cos x$, $y = \sin x$, and $y = \tan x$.
General form: $y - k = a \cos b(x - h)$, $y - k = a \sin b(x - h)$, and $y - k = a \tan b(x - h)$.

7.   a.

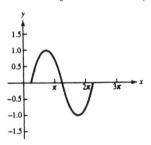

b.

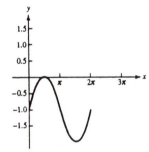

9.   a.

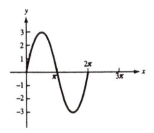

b.

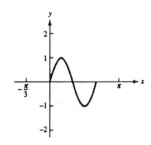

11. a.

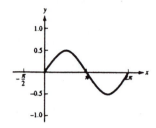

b.

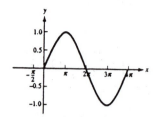

13.

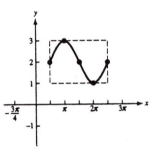

15.

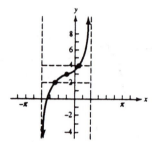

17.

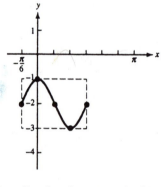

19. Amplitude of 4. Period of 2π. Possible equation with $(h, k) = (0, 0)$ is $y = 4\sin x$.

21. Amplitude of 2. Period of 2π. Possible equation with $(h, k) = (0, 0)$ is $y = 2\cos(x - \pi)$.

23. Amplitude of 10. Period of 2. Possible equation with $(h, k) = (0, 0)$ is $y - 20 = 10\sin \pi x$.

25. Amplitude of $\frac{1}{2}$. Period of 2π.

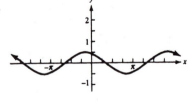

27. Amplitude of 2. Period of 1.

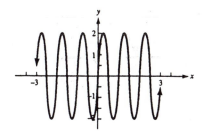

29. No amplitude, but $a = 4$. Period of 5.

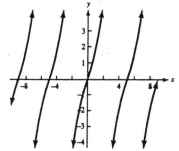

31. Amplitude of 1. Period of $\frac{\pi}{2}$.

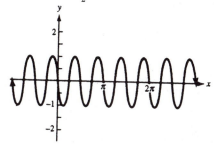

33. No amplitude, but $a = 2$. Period of $\frac{\pi}{2}$.

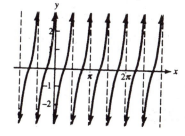

35. Amplitude of 2. Period of $\frac{2\pi}{3}$.

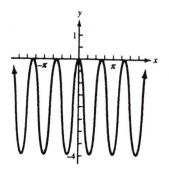

37. Amplitude of $\sqrt{2}$. Period of 2π.

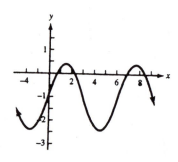

39. No amplitude, but $a = 2$. Period is 2π.

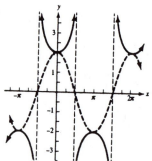

41. No amplitude, but $a = \frac{1}{2}$. Period is π.

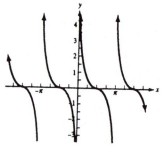

43. No amplitude, but $a = 1$. Period is π.

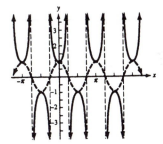

45. No amplitude, but $a = 2$. Period is 2π.

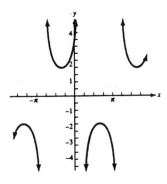

47. a. Amplitude is 20 mm of mercury.

  b. Pulse rate $= \frac{60}{0.6} = 100$ beats/min

49. Period is the reciprocal of frequency, thus $p = \frac{1}{f} = \frac{1}{330}$,
$b = \frac{2\pi}{b} = \frac{2\pi}{\frac{1}{330}} = 660\pi$, and $y = 0.02\cos(660\pi x)$.

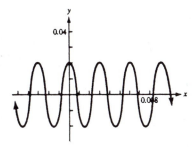

51. First plot the points and draw a smooth curve connecting them. Note the graph reaches a high point at $(15, 60)$. This appears to be a shifted sine or cosine curve with an amplitude of 60. Since the time elapsed between the starting low point and the high point is 15 seconds, the period must be 30 seconds. So, period $= \frac{2\pi}{b} = 30$ thus $b = \frac{\pi}{15}$. Also, from the data points, it appears that the current hits zero at 7.5 seconds. So, consider $(7.5, 0)$ to be the starting point or (h, k) of the sine curve. Finally, $I = 60\sin\frac{\pi}{15}(t - 7.5)$, where $0 \le t \le 120$.

53. First plot the points and draw a smooth curve connecting them. It appear to have an amplitude of 3 and a period of 1.2. Also, consider $(0.3, 0)$ to be the starting point of the sine curve. Thus, $y = 3\sin\frac{5\pi}{3}(x - 0.3)$ is an appropriate equation.

57. a.

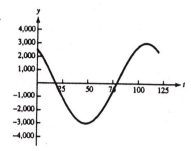

b.   3000 miles is the greatest distance reached north of the equator since 3000 is the amplitude.

c. $t = 2$ h (or 120 min).

Problem Set 3.3

3. $x \approx 2.5$ and $x \approx 0$

5. $x \approx 4.4$

7. With $0 \le x \le 2\pi$, $x \approx 1.1$ and $x \approx 5.3$.

9. $x \approx 0.8$

11. $x \approx 0.08$ and $x \approx 2.4$

13. $x \approx 0.5$ and $x \approx 3.7$

15.

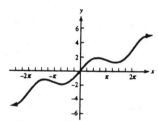

17.

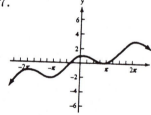

19.

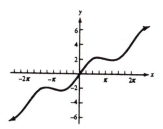

21.

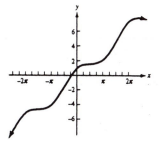

23.

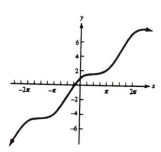

25.

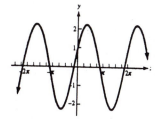

27.

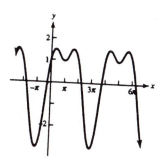

29.

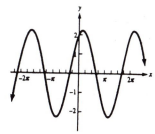

31.

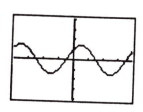

33.

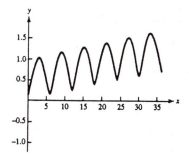

35. Sun curve: $y = \cos \frac{\pi}{6}x$; moon curve : $y = 4 \cos \frac{\pi}{6}x$, so the combined curve = sun curve + moon curve $= 5 \cos \frac{\pi}{6}x$.

37.

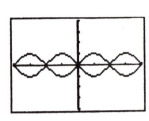

39.

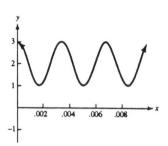

41. The frequency, $n = 10$ waves/min $= \frac{1}{6}$ waves/sec. The period, p, is the reciprocal of the frequency, so $p = 6$. Thus $\lambda = 5.09p^2 = (5.09)(6)^2 = 183.24$, and the phase velocity, $v = \frac{\lambda}{p} = \frac{183.24}{6} = 30.54$ ft/s. Now, convert the phase velocity to mi/h, thus $v = (30.54)(\frac{60^2}{5280}) = 20.82$ mi/h. The equation for the wave is $y = 3\sin 0.03(x - 30.54t)$.

43. $a = 12$ so $2a = 24$ ft from trough to crest; wavelength $\lambda \approx \frac{2\pi}{1} \approx 6.28$; $2\pi n = 37.7$, so $n = 6$; phase velocity is $v \approx 37.7$ ft/s ≈ 25.7 mi/h.

45.

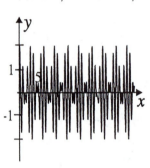

47.

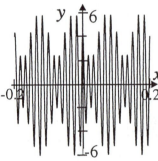

49. a.

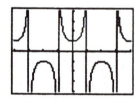

b.

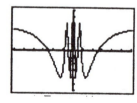

51.

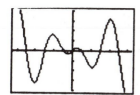

53.

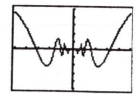

55.

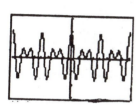

57.

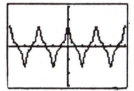

59.

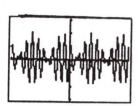

61.

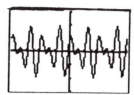

63.

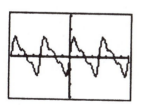

65. a. $m = \frac{1}{2}$ b. $m = \frac{1}{4}$ c. $m = \frac{1}{8}$ d. Answers vary

Problem Set 3.4

1. The range for the inverse cosine is $0 \le y \le \pi$.

3. quadrants I, IV

5. The range for the inverse tangent is $-\frac{\pi}{2} \le y \le \frac{\pi}{2}$.

13. a. 0.7   b. 0.0   c. -1.0 d. -0.7

      e. 0.0   f. $\frac{\pi}{2}$ g. $\frac{\pi}{3}$ h. π

      i.

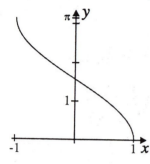

15. a. 1.1   b. 0.4   c. 0.0   d. -1.0

      e. 0.8   f. 1.1   g. 0.5   h. -0.8

      i.

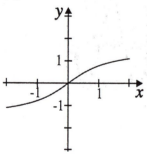

17. a. $\frac{\pi}{4}$ b. $\cot^{-1} 1 = \tan^{-1} 1 = \frac{\pi}{4}$

      c. $-\frac{\pi}{2}$ or $\frac{3\pi}{2}$ d. 0

19. a. $\frac{\pi}{3}$ b. $-\frac{\pi}{2}$ or $\frac{3\pi}{2}$ c. $-\frac{\pi}{3}$ d. $\frac{3\pi}{4}$

21. a. $-\frac{\pi}{6}$ b. 0   c. $\frac{\pi}{3}$ d. $\frac{\pi}{3}$

23. a. -1.31 b. 2.81   c. 1.11   d. 0.98

25. a. $21°$ b. $19°$ c. $46°$ d. $-28°$

27. a. $153°$ b. $15°$ c. $100°$ d. $-72°$

29. True

31. False, it is in quadrant II.

33. True

35. True

37. True

39. False, since $\cos^{-1} x \neq \sec^{-1} x$.

41. a. $\frac{1}{3}$ b. $\frac{\pi}{15}$

43. a. 1 b. $\frac{\pi}{6}$

45. a. $\frac{1}{2}$ b. $\frac{\sqrt{5}}{3}$

47. Let $\theta = \sin^{-1} x$ so that $\sin \theta = x$. Thus, $\cos(\sin^{-1} x) = \cos \theta = \pm\sqrt{1 - \sin^2 \theta} = \pm\sqrt{1 - x^2}$.

49. Let $\theta = \sec^{-1} x$ so that $\sec \theta = x$. Thus, $\cos(\sec^{-1} x) = \cos \theta = \dfrac{1}{\sec \theta} = \dfrac{1}{x}$.

51. Let $\theta = \sec^{-1} x$ so that $\sec \theta = x$. Thus,

$$\tan(\sec^{-1} x) = \tan \theta = \frac{\sin \theta}{\cos \theta} = \sec \theta (\pm\sqrt{1 - \cos^2 \theta}) = \left(\pm\sqrt{1 - \frac{1}{\sec^2 \theta}}\right) x = \left(\pm\sqrt{1 - \frac{1}{x^2}}\right) x$$

53. $\tan^{-1}\left(\frac{5}{8}\right) \approx 32.005°$, make the parking angle $32°$.

55. 1.50

57. $L = 62$ in.

59.

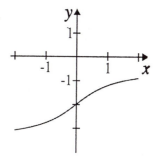

61.

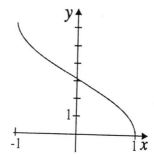

63.

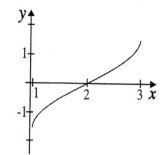

65.

$$\sec \theta = x \qquad \text{Definition}$$

$$\frac{1}{\cos \theta} = x \qquad \text{Reciprocal identity}$$

$$\cos \theta = \frac{1}{x} \qquad \text{Solve for cosine } \theta$$

$$\cos^{-1} \frac{1}{x} = \theta \qquad \text{Definition}$$

67. $\theta = \cot^{-1} x \qquad \dfrac{\pi}{2} \le \theta \le \pi$

$x = \cot \theta \qquad \dfrac{\pi}{2} \le \theta \le \pi$

$\tan \theta = \dfrac{1}{x}$

$\tan(\theta - \pi) = \dfrac{1}{x} \qquad -\dfrac{\pi}{2} \le \theta - \pi \le 0$

$\theta - \pi = \tan^{-1} \dfrac{1}{x}$

$\cot^{-1} \theta - \pi = \tan^{-1} \dfrac{1}{x} \qquad \text{Substitute in for } \theta \text{ from step 1.}$

$\cot^{-1} \theta = \tan^{-1} \dfrac{1}{x} + \pi$

Chapter 3 Sample Test

1. a. $y = \tan x$ b. $y = \sec x$ c. $y = \cos x$

     d. $y = \cot x$ e. $y = \csc x$ f. $y = \sin x$

3. a. $-\dfrac{\pi}{4}$ b. $\dfrac{2\pi}{3}$ c. $\dfrac{3\pi}{4}$ d. $\dfrac{2\pi}{3}$

     e. 0.848 f. 1.318 g. 1.117 h. -1.309

5. a. (h, k) b. $2|a|$ c. $\dfrac{2\pi}{b}$ d. $\dfrac{\pi}{b}$

7.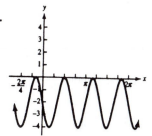

9.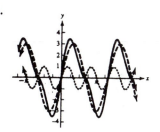

Chapter 3 Miscellaneous Problems

3. a. 0 to 1   b. decreases   c. decreases   d. increases

5. a. $\theta = \frac{2}{\sqrt{13}}$ b. $\theta = \frac{\sqrt{21}}{5}$

7. a. $(-4\pi, 0)$ b. $(-\pi, 0)$ c. $(\pi/2, 0)$ d. $(2\pi, 0)$ e. $(7\pi/2, -1)$

9. If $\cos\beta = \dfrac{x}{r}$, $\sin\beta = \dfrac{y}{r}$, and $r = 5$, then $x = 5\cos\beta$ and $y = 5\sin\beta$. So the point of intersection is $(x, y) = (5\cos\beta, 5\sin\beta)$.

11. $y = \frac{1}{2}\sin\frac{1}{2}x$

13. $y = 5\cos\dfrac{\pi}{5}x$

15.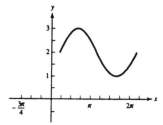

17.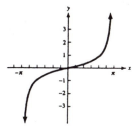

19.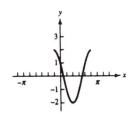

21.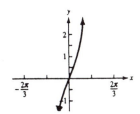

23.

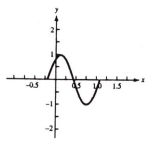

25.

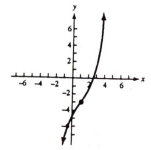

27.

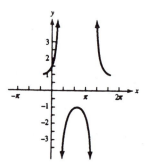

29.

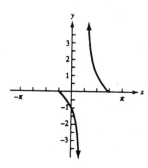

31. a.

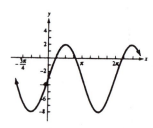

b.

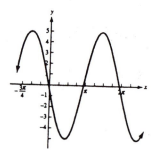

33.

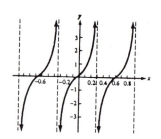

35.

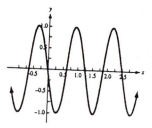

37.

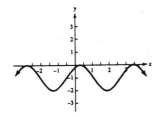

39.

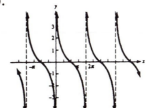

41.

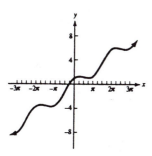

43.

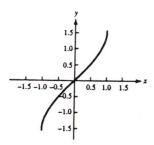

45.

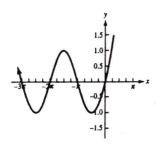

47.

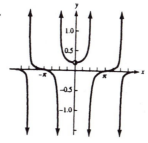

49.

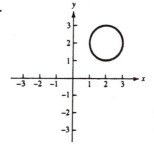

51.

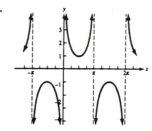

53. With $0 \le x < 2\pi$, $x \approx 0.7$ and $x \approx 2.5$.

55. $2x = 0.5 \sin x$ so $4x = \sin x$ thus $x = 0$ is the only solution.

57.

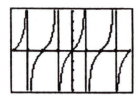

59. Change all measurements to ft/s. So, $v = 300 \text{ mi/h} = 440 \text{ ft/s}$;
$\lambda = 120 \text{ mi} = 633600 \text{ ft}$; $p = 30 \text{ min} = 1800 \text{ sec}$;
$y = 2\sin \dfrac{2\pi}{633600}(x - 440t)$. At $t = 0$, the equation becomes
$y = 2\sin \dfrac{2\pi}{633600}(x)$.

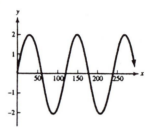

61. To calculate how long it will take the north pole to drift $1°$, first convert radians per year to degrees per year. Thus, $(0.0017 \text{ rad/yr })(\frac{180°}{\pi}) = 0.0974 \text{ deg/century}$, and $1° = (x)(0.0974 \text{ deg/yr})$ which will take about 1,000 years.

Chapters 1–3 Cumulative Review

3. a. $\frac{\pi}{6}$ b. $\frac{\pi}{4}$ c. $\frac{\pi}{3}$ d. $\frac{\pi}{2}$ e. π f. $\frac{3\pi}{2}$ g. 2π
  h. $\frac{2\pi}{3}$ i. $\frac{3\pi}{4}$ j. $\frac{5\pi}{6}$ k. $-\frac{\pi}{2}$ l. $-\frac{5\pi}{3}$ m. $-\frac{4\pi}{3}$ n. $-\frac{5\pi}{4}$

9. a. secant   b. tangent   c. cosecant   d. cosine   e. sine   f. cotangent

11. $\sin^2\theta + \cos^2\theta = 1, \tan^2\theta + 1 = \sec^2\theta, 1 + \cot^2\theta = \csc^2\theta$

13. a. 1   b. 1   c. $\frac{\sqrt{3}}{2}$
  d. 1   e. 0   f. $\frac{\sqrt{2}}{2}$

15. a. $\frac{1}{2}$ b. $\frac{1}{2}$ c. 0
  d. 1   e. undefined   f. undefined

17. a. 2   b. $\sqrt{3}$ c. 0
  d. -1 e. $\sqrt{3}$ f. 0

19. a. undefined   b. -1 c. undefined

d. $\frac{2}{\sqrt{3}}$ e. $\frac{\sqrt{3}}{3}$ f. undefined

21. a. $\frac{\pi}{6}$ b. $\frac{\pi}{6}$ c. 0

      d. $\frac{\pi}{4}$ e. $\frac{\pi}{2}$ f. $\frac{\pi}{3}$

23.

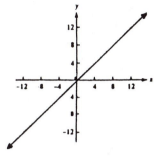

25.

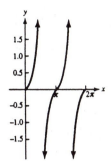

27.

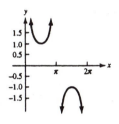

29.

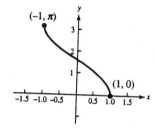

31.

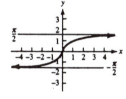

39. A   **41.** C   **43.** A   **45.** C   **47.** A   **49.** C

51. D   **53.** B   **55.** E

57. $\cos \alpha = \frac{\sqrt{7}}{4}$; $\sin \alpha = \frac{3}{4}$; $\tan \alpha = \frac{3}{\sqrt{7}}$; $\sec \alpha = \frac{4}{\sqrt{7}}$; $\csc \alpha \frac{4}{3}$; $\cot \alpha = \frac{\sqrt{7}}{3}$

59. $a = 15.24, b = 23.49, c = 28.0, \alpha = 32.98°, \beta = 57.02°, \gamma = 90°$

61. $a = 7.41, b = 5.37, c = 9.15, \alpha = 35.9°, \beta = 54.1°, \gamma = 90°$

63. a.

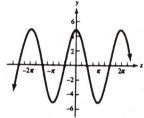

b.

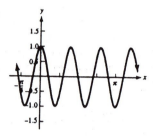

65. a.

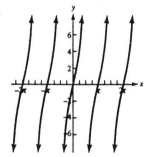

b.

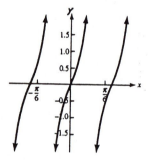

67.

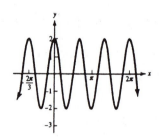

69.

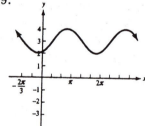

71.

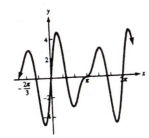

73. We can first find the angle between the two cities that converges at the UFO, since $\theta = 180° - 11.9° - 35.8° = 132.3°$. Then we can use the Law of Sines to determine the distance between the UFO and one of the cities, since $\dfrac{\sin 132.3°}{12.6} = \dfrac{\sin 11.9°}{y}$, and thus $y = 3.513$ mi. Then we can use the Law of Sines again to find the height of the UFO, since $\dfrac{h}{\sin 35.8°} = \dfrac{3.513}{\sin 90°}$, and thus $h = 2.005$ mi. Converting this to ft gives $h = 10850 \approx 10900$ ft.

75. Middle gear is turning at $20\left(\frac{5}{3}\right) = 33\frac{1}{2}$ rev/s and the small gear is turning at $20\left(\frac{5}{3}\right)\left(\frac{3}{1}\right) = 100$ rev/s.

77. a.

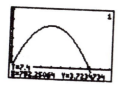

b. At the time of impact $y = 0$, so the equation for y is $y = 160\sin 48°t - 16t^2 = 118.9031721t - 16t^2$. Then $0 = 118.9031721t - 16t^2$ gives $t \approx 7.4$.

c. The equation for x to find the horizontal distance is $x = 160\cos 48°t - 16t^2$. Then using $t = 7.4$ from part (b), $x = 107.060897(7.4) \approx 792$ ft.

CHAPTER 4   TRIGONOMETRIC EQUATIONS AND IDENTITIES

Problem Set 4.1

5. a. $x \approx 1.05$ b. $x = \frac{\pi}{3}$

    c. $x = \frac{\pi}{3} + 2k\pi; \frac{5\pi}{3} + 2k\pi$

7. a. $x \approx 2.09$ b. $x = \frac{2\pi}{3}; \frac{4\pi}{3}$

    c. $x = \frac{2\pi}{3} + 2k\pi; \frac{4\pi}{3} + 2k\pi$

9. a. $x \approx 0.79$ b. $x = \frac{\pi}{4}; \frac{5\pi}{4}$

    c. $x = \frac{\pi}{4} + k\pi$

11. a. $x \approx 2.09$ b. $x = \frac{2\pi}{3}$

    c. $x = \frac{2\pi}{3} + 2k\pi; \frac{5\pi}{3} + 2k\pi$

13. a. $x \approx 0.79$ b. $x = \frac{\pi}{4}; \frac{3\pi}{4}$

    c. $x = \frac{\pi}{4} + 2k\pi; \frac{3\pi}{4} + 2k\pi$

15. a. $x \approx -0.52$ b. $x = \frac{2\pi}{3}; \frac{5\pi}{6}; \frac{5\pi}{3}; \frac{11\pi}{6}$

    c. $x = \frac{2\pi}{3} + 2k\pi; \frac{5\pi}{6} + 2k\pi; \frac{5\pi}{3} + 2k\pi; \frac{11\pi}{6} + 2k\pi$

17. a. $x \approx 0.26$ b. $x = \frac{\pi}{12}; \frac{\pi}{4}; \frac{3\pi}{4}; \frac{11\pi}{12}; \frac{17\pi}{12}; \frac{19\pi}{12}$

    c. $x = \frac{\pi}{12} + 2k\pi; \frac{\pi}{4} + 2k\pi; \frac{3\pi}{4} + 2k\pi; \frac{11\pi}{12} + 2k\pi; \frac{17\pi}{12} + 2k\pi; \frac{19\pi}{12} + 2k\pi$

19. a. $x \approx -0.39$ b. $x = \frac{5\pi}{8}; \frac{7\pi}{8}; \frac{13\pi}{8}; \frac{15\pi}{8}$

    c. $x = \frac{5\pi}{8} + k\pi; \frac{7\pi}{8} + k\pi$

21. a. $x \approx -0.26$ b. $x = \frac{\pi}{4}; \frac{7\pi}{12}; \frac{11\pi}{12}; \frac{5\pi}{4}; \frac{19\pi}{12}; \frac{23\pi}{12}$

    c. $x = \frac{\pi}{4} + k\pi; \frac{7\pi}{12} + k\pi; \frac{11\pi}{12} + k\pi$

23. a. $x \approx -0.39$ b. $x = \frac{3\pi}{8}; \frac{7\pi}{8}; \frac{11\pi}{8}; \frac{15\pi}{8}$

    c. $x = \frac{3\pi}{8} + k\pi; \frac{7\pi}{8} + k\pi$

25. a. $x \approx 0.00$ b. $x = 0; \pi$ c. $x = 0 + k\pi$

27. a. $x = 0.00$ b. $x = 0; \frac{\pi}{2}; \pi; \frac{3\pi}{2}$ c. $x = 0 + \frac{k\pi}{2}$

29. a. $x \approx 1.05; 0.52$ b. $x = \frac{\pi}{3}; \frac{\pi}{6}; \frac{5\pi}{3}; \frac{5\pi}{6}$

    c. $x = \frac{\pi}{3} + 2k\pi; \frac{\pi}{6} + 2k\pi; \frac{5\pi}{2} + 2k\pi; \frac{5\pi}{6} + 2k\pi$

31.

$$\tan^2 x = \sqrt{3}\tan x$$
$$\tan x = \sqrt{3}$$
$$x = 0; \pi; \frac{\pi}{3}; \frac{4\pi}{3}$$

33.

$$\cos^2 x = \frac{1}{2}$$
$$\cos x = \frac{1}{\sqrt{2}}$$
$$x = \frac{\pi}{4}; \frac{3\pi}{5}; \frac{5\pi}{4}; \frac{7\pi}{4}$$

35.

$$\sqrt{2}\cos x \sin x = \sin x$$
$$\cos x = \frac{1}{\sqrt{2}}$$
$$x = 0; \pi; \frac{\pi}{4}; \frac{7\pi}{4}$$

37. $x = \frac{5\pi}{6} + 2k\pi; \frac{7\pi}{6} + 2k\pi$

39. $x \approx 0.40 + 2k\pi; 2.74 + 2k\pi$

41. $x \approx 0.94 + k\pi$

43. Factoring gives $(\sin x - 2)(\sin x + 1) = 0$ and $x = \frac{3\pi}{2} \approx 4.71$.

45. Using the quadratic formula, $\csc x = \dfrac{1 \pm \sqrt{5}}{2}$ and $x \approx 0.67$.

47. Dividing both sides by $2\sin 2x$ gives $\cos 3x = \frac{1}{2}$ and $x \approx 0.35$.

49. Factoring $\sin 2x$ from both terms gives $\sin 2x(1 + 2\cos x) = 0$ and $x \approx 0, x \approx 2.09$.

51. $x \approx 0.38$

53. $x \approx 0.32$

55. Factoring gives $(-2\sin x + 1)(\sin x + 1) = 0$ and $x \approx 0.52, x \approx -1.57$.

57. Factoring $\cos x$ from both terms gives $\cos x(2\sin x + 1) = 0$ and $x \approx 1.57, x \approx -0.52$.

59. Divide both sides by $2\cos x$ and $x \approx 0.52, x \approx 1.57$.

61.
$$\sin^2 3x + \sin 3x = 1 - \sin^2 3x + 1$$
$$2(\sin^2 3x) + \sin 3x - 2 = 0$$
$$\sin 3x = -\frac{1}{4} + \frac{1}{4}\sqrt{17}$$
$$x \approx 0.30$$

63. $t \approx 0.185$

65.
$$\cos 2x - 1 = \sin 2x$$
$$\frac{\cos 2x}{\sin 2x} = \frac{1}{\sin 2x}$$
$$\cot 2x = \csc 2x$$
$$x \approx 0, 2.4; 3.1, 5.5$$

67. none

69.
$$\frac{1}{\sin \frac{x}{3}} + \frac{1}{\cos \frac{x}{2}} = \frac{\pi}{3}$$
$$1 + \frac{\sin \frac{x}{3}}{\cos \frac{x}{3}} = \frac{\pi}{3} \sin \frac{x}{3}$$
$$\cos \frac{x}{2} + \sin \frac{x}{3} = \frac{\pi}{3} \sin \frac{x}{3} \cos \frac{x}{2}$$
$$x \approx 1.9$$

71.
$$\cos^2 x \cos x - (1 - \cos^2 x)\sin x = \frac{1}{2}$$
$$\cos^2 x \cos x - \sin x + \cos^2 x \sin x = \frac{1}{2}$$
$$\cos^2 x (\cos x + \sin x) - \sin x = \frac{1}{2}$$
$$x \approx 0.5, 4.2$$

Problem Set 4.2

7. $\sin\theta = -3u$ and $\csc\theta = -\frac{1}{3u}$; $\cos\theta = \sqrt{1-(-3u)^2} = \sqrt{1-9u^2}$ and $\sec\theta = \dfrac{1}{\sqrt{1-9u^2}}$;

$\tan\theta = -\dfrac{3u}{\sqrt{1-9u^2}}$ and $\cot\theta = -\dfrac{\sqrt{1-9u^2}}{3u}$

9. $\cos\theta = -\frac{2u}{5}$ and $\sec\theta = -\frac{5}{2u}$; $\sin\theta = \sqrt{1-\frac{4u^2}{25}} = \dfrac{\sqrt{25-u^2}}{5}$ and $\csc\theta = \dfrac{5}{\sqrt{25-4u^2}}$; $\tan\theta =$

$-\dfrac{\sqrt{25-4u^2}}{2u}$ and $\cot\theta = -\dfrac{2u}{\sqrt{25-4u^2}}$

11.

$$\cot\theta\tan^2\theta = \frac{\tan^2\theta}{\tan\theta}$$
$$= \tan\theta$$

13.

$$\tan^2\theta - \sin^2\theta = \frac{\sin^2\theta}{\cos^2\theta} - \sin^2\theta\cdot\frac{\cos^2\theta}{\cos^2\theta}$$
$$= \frac{\sin^2\theta - \sin^2\theta\cos^2\theta}{\cos^2\theta}$$
$$= \sin^2\theta\frac{(1-\cos^2\theta)}{\cos^2\theta}$$
$$= \sin^2\theta\frac{\sin^2\theta}{\cos^2\theta}$$
$$= \sin^2\theta\tan^2\theta$$
$$= \tan^2\theta\sin^2\theta$$

15.

$$\tan\theta + \cot\theta = \frac{\sin\theta}{\cos\theta}\cdot\frac{\sin\theta}{\sin\theta} + \frac{\cos\theta}{\sin\theta}\cdot\frac{\cos\theta}{\cos\theta}$$
$$= \frac{\sin^2\theta + \cos^2\theta}{\cos\theta\sin\theta}$$
$$= \frac{1}{\cos\theta\sin\theta}$$
$$= \sec\theta\csc\theta$$

17.

$$\frac{1 - \sec^2 \beta}{\sec^2 \beta} = \cos^2 \beta (1 - \sec^2 \beta)$$

$$= \cos^2 \beta [1 - (1 + \tan^2 \beta)]$$

$$= \cos^2 \beta [- \tan^2 \beta]$$

$$= - \cos^2 \beta \frac{\sin^2 \beta}{\cos^2 \beta}$$

$$= - \sin^2 \beta$$

19.

$$\frac{1 - \sin^2 2\theta}{1 + \sin 2\theta} = \frac{(1 + \sin 2\theta)(1 - \sin 2\theta)}{(1 + \sin 2\theta)}$$

$$= 1 - \sin 2\theta$$

21.

$$\frac{1 + \cos 2\lambda \sec 2\lambda}{\tan 2\lambda + \sec 2\lambda} = \frac{1 + (\cos 2\lambda)\left(\frac{1}{\cos 2\lambda}\right)}{\tan 2\lambda + \sec 2\lambda}$$

$$= \frac{2}{\tan 2\lambda + \sec 2\lambda}\left(\frac{\cos 2\lambda}{\cos 2\lambda}\right)$$

$$= \frac{2 \cos 2\lambda}{\sin 2\lambda + 1}$$

23.

$$(\sin \alpha + \cos \alpha)^2 + (\sin \alpha - \cos \alpha)^2 = \sin^2 \alpha + 2 \sin \alpha \cos \alpha + \cos^2 \alpha$$

$$+ \sin^2 \alpha - 2 \sin \alpha \cos \alpha + \cos^2 \alpha$$

$$= (\sin^2 \alpha + \cos^2 \alpha) + (\sin^2 \alpha + \cos^2 \alpha)$$

$$= 1 + 1 = 2$$

25.

$$\frac{1 + \cot^2 \gamma}{1 + \tan^2 \gamma} = \frac{\csc^2 \gamma}{\sec^2 \gamma}$$

$$= \frac{\cos^2 \gamma}{\sin^2 \gamma}$$

$$= \cot^2 \gamma$$

27.

$$\tan^2 \phi + \sin^2 \phi + \cos^2 \phi = \tan^2 \phi + 1$$
$$= \sec^2 \phi$$

29.

$$2 - \cos^2 \theta = 2 - (1 - \sin^2 \theta)$$
$$= 1 + \sin^2 \theta$$

31.

$$\frac{1}{1 + \cos 2\theta} + \frac{1}{1 - \cos 2\theta} = \frac{1}{1 + \cos 2\theta}\left(\frac{1 - \cos 2\theta}{1 - \cos 2\theta}\right) + \frac{1}{1 - \cos 2\theta}\left(\frac{1 + \cos 2\theta}{1 + \cos 2\theta}\right)$$
$$= \frac{1 - \cos 2\theta + 1 + \cos 2\theta}{1 - \cos^2 2\theta}$$
$$= 2\csc^2 2\theta$$

33.

$$\frac{1}{\tan 2\beta} + \frac{\cos 2\beta}{\cot 2\beta} = \frac{\cos 2\beta}{\sin 2\beta} + \cos 2\beta \cdot \frac{\sin 2\beta}{\cos 2\beta}$$
$$= \cot 2\beta + \sin 2\beta$$

35. If we let $\theta = \frac{\pi}{4}$, then

$$2\cos 2\left(\frac{\pi}{4}\right)\sin 2\left(\frac{\pi}{4}\right) = \sin 2\left(\frac{\pi}{4}\right)$$
$$0 \neq 1$$

37. If we let $\theta = 0$, then

$$\cos(0) - 3\sin(0) + 3 = 0$$
$$4 \neq 0$$

39. If we let $\theta = 0$, then

$$2\sin^2(0) - 2\cos^2(0) = 1$$
$$-2 \neq 1$$

41.

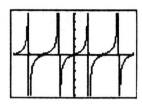

43.

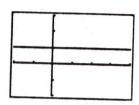

45. False, you can't cancel out the 2 inside $\sin 2\theta$.

47.

$$\frac{1 + \tan\alpha}{1 - \tan\alpha} = \frac{(1 + \tan\alpha)}{(1 - \tan\alpha)} \cdot \frac{(1 + \tan\alpha)}{(1 + \tan\alpha)}$$

$$= \frac{1 + 2\tan\alpha + \tan^2\alpha}{1 - \tan^2\alpha}$$

$$= \frac{(1 + \tan^2\alpha) + 2\tan\alpha}{1 - (\sec^2\alpha - 1)}$$

$$= \frac{\sec^2\alpha + 2\tan\alpha}{2 - \sec^2\alpha}$$

49.

$$\frac{\sin^3 x - \cos^3 x}{\sin x - \cos x} = \frac{(\sin x - \cos x)(\sin^2 x + \sin x \cos x + \cos^2 x)}{(\sin x - \cos x)}$$

$$= \sin^2 x + \sin x \cos x + \cos^2 x$$

$$= (\sin^2 x + \cos^2 x) + \sin x \cos x$$

$$= 1 + \sin x \cos x$$

51.

$$\frac{(\sec^2\gamma+\tan^2\gamma)^2}{\sec^4\gamma-\tan^4\gamma} = \frac{(\sec^2\gamma+\tan^2\gamma)^2}{(\sec^2\gamma-\tan^2\gamma)(\sec^2\gamma+\tan^2\gamma)}$$

$$= \frac{\sec^2\gamma+\tan^2\gamma}{\sec^2\gamma-\tan^2\gamma}$$

$$= \frac{\sec^2\gamma+\tan^2\gamma}{1+\tan^2\gamma-\tan^2\gamma}$$

$$= \sec^2\gamma+\tan^2\gamma$$

$$= 1+\tan^2\gamma+\tan^2\gamma$$

$$= 1+2\tan^2\gamma$$

53.

$$\frac{1}{\sec\theta+\tan\theta} = \frac{1}{\sec\theta+\tan\theta}\cdot\frac{\sec\theta-\tan\theta}{\sec\theta-\tan\theta}$$

$$= \frac{\sec\theta-\tan\theta}{\sec^2\theta-\tan^2\theta}$$

$$= \frac{\sec\theta-\tan\theta}{1+\tan^2\theta-\tan^2\theta}$$

$$= \sec\theta-\tan\theta$$

55.

$$\frac{1-\sec^3\theta}{1-\sec\theta} = \frac{(1-\sec\theta)(1+\sec\theta+\sec^2\theta)}{1-\sec\theta}$$

$$= 1+\sec\theta+\tan^2\theta+1$$

$$= \tan^2\theta+\sec\theta+2$$

57.

$$\sqrt{(3\cos\theta-4\sin\theta)^2+(3\sin\theta+4\cos\theta)^2}$$

$$= \sqrt{9\cos^2\theta-24\cos\theta\sin\theta+16\sin^2\theta+24\sin\theta\cos\theta+16\cos^2\theta}$$

$$= \sqrt{25\cos^2\theta+25\sin^2\theta}$$

$$= \sqrt{25(\cos^2\theta+\sin^2\theta)}$$

$$= \sqrt{25}=5$$

59.

$$\frac{\cos^4\theta - \sin^4\theta}{(\cos^2\theta - \sin^2\theta)^2} = \frac{(\cos^2\theta - \sin^2\theta)(\cos^2\theta + \sin^2\theta)}{(\cos^2\theta - \sin^2\theta)^2}$$

$$= \frac{1}{\cos^2\theta - \sin^2\theta}$$

61.

$$(\sec\alpha + \sec\beta)^2 - (\tan\alpha - \tan\beta)^2$$
$$= \sec^2\alpha + 2\sec\alpha\sec\beta + \sec^2\beta - \tan^2\alpha + 2\tan\alpha\tan\beta + \tan^2\beta$$
$$= \tan^2\alpha + 1 + 2\sec\alpha\sec\beta + \tan^2\beta + 1 - \tan^2\alpha + 2\tan\alpha\tan\beta - \tan^2\beta$$
$$= 2 + 2(\sec\alpha\sec\beta + \tan\alpha\tan\beta)$$

63.

$$\tan\theta + \sec\theta + 1 = (\tan\theta + \sec\theta + 1)\left(\frac{\tan\theta + \sec\theta - 1}{\tan\theta + \sec\theta - 1}\right)$$

$$= (\tan^2\theta + \tan\theta\sec\theta - \tan\theta + \sec\theta\tan\theta + \sec^2\theta - \sec\theta + \tan\theta$$
$$+ \sec\theta - 1)\left(\frac{1}{\tan\theta + \sec\theta - 1}\right)$$

$$= \frac{\tan^2\theta + 2\tan\theta\sec\theta + (\sec^2\theta - 1)}{\tan\theta + \sec\theta - 1}$$

$$= \frac{2\tan^2\theta + 2\tan\theta\sec\theta}{\tan\theta + \sec\theta - 1}$$

65. $\sqrt{4 - u^2} = \sqrt{4 - (2\sin x)^2} = \sqrt{4(1 - \sin^2 x)} = 2\cos x$

67. $\sqrt{1 + u^2} = \sqrt{1 + \tan^2 x} = \sqrt{\sec^2 x} = \sec x$

69. When $x = 100$, $y = 1$ and since $-1 < \sin x < 1$ then there will be no more intersections and this also true for $x = -100$. Thus, each repetition of $\sin x$ gives two intersections with 32 on the positive x-axis and 32 on the negative x-axis. So $32 + 32 = 64$ solutions.

Problem Set 4.3

1. The distance formula is $d = \sqrt{(x_2 - x_1)^2 + (y_2 - y_1)^2}$.

3. The opposite angle identities are $\cos(-\theta) = \cos\theta$, $\sin(-\theta) = -\sin\theta$, and $\tan(-\theta) = -\tan\theta$.

5. $\sin(30° + 45°) \approx 0.9664$ and $\sin 30° + \sin 45° \approx 1.207$.

7. $\sin\left(\frac{\pi}{2} - \theta\right) = \cos\theta \neq 1 - \sin\theta$

9. $\sin(\theta + 45°) = \sin\theta\cos 45° + \cos\theta\sin 45° = \frac{\sqrt{2}}{2}(\sin\theta + \cos\theta)$

11. $\cos(\theta - 45°) = \cos\theta\cos 45° + \sin\theta\sin 45° = \frac{\sqrt{2}}{2}(\cos\theta + \sin\theta)$

13. $\tan(45° + \theta) = \dfrac{\tan 45° + \tan\theta}{1 - \tan 45°\tan\theta} = \dfrac{1 + \tan\theta}{1 - \tan\theta}$

15. $\sin(\theta + \theta) = \sin\theta\cos\theta + \cos\theta\sin\theta = 2\sin\theta\cos\theta$

17. $\sin 15° = \sin(90° - 15°) = \cos 75°$

19. $\tan 62° = \tan(90° - 62°) = \cot 28°$

21. $\cos\dfrac{5\pi}{6} = \cos(\dfrac{\pi}{2} - \dfrac{5\pi}{6}) = \sin\left(-\dfrac{\pi}{3}\right) = -\sin\dfrac{\pi}{3}$

23. $\sin(-23°) = -\sin(23°)$

25. $\cos(-57°) = \cos(57°)$

27. $\tan(-29°) = -\tan(29°)$

29. 0.9135

31. 0.9511

33. 1.1918

35.

$$\sin 15° = \sin(45° - 30°)$$
$$= \sin 45° \cos 30° - \cos 45° \sin 30°$$
$$= \frac{\sqrt{2}}{2}\left(\frac{\sqrt{3}}{2}\right) - \frac{\sqrt{2}}{2}\left(\frac{1}{2}\right)$$
$$= \frac{\sqrt{6} - \sqrt{2}}{4}$$

$$\cos 15° = \cos(45° - 30°)$$
$$= \cos 45° \cos 30° + \sin 45° \sin 30°$$
$$= \frac{\sqrt{2}}{2}\left(\frac{\sqrt{3}}{2}\right) + \frac{\sqrt{2}}{2}\left(\frac{1}{2}\right)$$
$$= \frac{\sqrt{6} + \sqrt{2}}{4}$$

$$\tan 15° = \tan(45° - 30°)$$
$$= \frac{\tan 45° - \tan 30°}{1 + \tan 45° \tan 30°}$$
$$= \frac{3 - \sqrt{3}}{3 + \sqrt{3}}$$

37.

$$\sin 75° = \sin(45° + 30°)$$
$$= \sin 45° \cos 30° + \cos 45° \sin 30°$$
$$= \frac{\sqrt{2}}{2}\left(\frac{\sqrt{3}}{2}\right) + \frac{\sqrt{2}}{2}\left(\frac{1}{2}\right)$$
$$= \frac{\sqrt{6} + \sqrt{2}}{4}$$

$$\cos 75° = \cos(45° + 30°)$$
$$= \cos 45° \cos 30° - \sin 45° \sin 30°$$
$$= \frac{\sqrt{2}}{2}\left(\frac{\sqrt{3}}{2}\right) - \frac{\sqrt{2}}{2}\left(\frac{1}{2}\right)$$
$$= \frac{\sqrt{6} - \sqrt{2}}{4}$$

$$\tan 75° = \tan(45° + 30°)$$
$$= \frac{\tan 45° + \tan 30°}{1 - \tan 45° \tan 30°}$$
$$= \frac{3 + \sqrt{3}}{3 - \sqrt{3}}$$

39.

$$\sin 165° = \sin(15°) = \sin(45° - 30°) \qquad \cos 165° = -\cos 15° = -\cos(45° - 30°)$$

$$= \sin 45° \cos 30° - \cos 45° \sin 30° \qquad = -[\cos 45° \cos 30° + \sin 45° \sin 30°]$$

$$= \frac{\sqrt{2}}{2}\left(\frac{\sqrt{3}}{2}\right) - \frac{\sqrt{2}}{2}\left(\frac{1}{2}\right) \qquad = -\left[\frac{\sqrt{2}}{2}\left(\frac{\sqrt{3}}{2}\right) + \frac{\sqrt{2}}{2}\left(\frac{1}{2}\right)\right]$$

$$= \frac{\sqrt{6} - \sqrt{2}}{4} \qquad\qquad = \frac{-\sqrt{6} - \sqrt{2}}{4}$$

$$\tan 165° = -\tan 15° = -\tan(45° - 30°)$$

$$= -\left[\frac{\tan 45° - \tan 30°}{1 + \tan 45° \tan 30°}\right]$$

$$= \frac{-3 + \sqrt{3}}{3 + \sqrt{3}}$$

41.   a. $\cos\left(\frac{\pi}{3} - \theta\right) = \cos\left[-\left(\theta - \frac{\pi}{3}\right)\right] = \cos\left(\theta - \frac{\pi}{3}\right)$

   b. $\sin\left(\frac{\pi}{3} - \theta\right) = \sin\left[-\left(\theta - \frac{\pi}{3}\right)\right] = -\sin\left(\theta - \frac{\pi}{3}\right)$

   c. $\tan\left(\frac{\pi}{3} - \theta\right) = \tan\left[-\left(\theta - \frac{\pi}{3}\right)\right] = -\tan\left(\theta - \frac{\pi}{3}\right)$

43.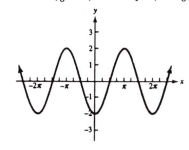

45.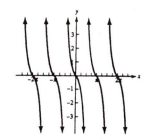

47.

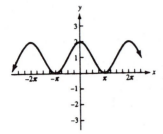

49. a. even   b. even   c. odd   d. even

51.

$$
\begin{aligned}
\tan(\alpha - \beta) &= \tan[\alpha + (-\beta)] \\
&= \frac{\tan \alpha + \tan(-\beta)}{1 - \tan \alpha \tan(-\beta)} \\
&= \frac{\tan \alpha - \tan \beta}{1 + \tan \alpha \tan \beta}
\end{aligned}
$$

53.

$$
\begin{aligned}
\frac{\sin 6\theta}{\sin 3\theta} - \frac{\cos 6\theta}{\cos 3\theta} &= \frac{\sin 6\theta \cos 3\theta - \cos 6\theta \sin 3\theta}{\sin 3\theta \cos 3\theta} \\
&= \frac{\sin(6\theta - 3\theta)}{\sin 3\theta \cos 3\theta} \\
&= \frac{1}{\cos 3\theta} \\
&= \sec 3\theta
\end{aligned}
$$

55.

$$
\begin{aligned}
\sin(\alpha - \beta) \cos \beta - \sin(\alpha - \beta) \sin \beta &= \cos \alpha \cos^2 \beta + \sin \alpha \sin \beta \cos \beta \\
&\quad - \sin \alpha \cos \beta \sin \beta + \cos \alpha \sin^2 \beta \\
&= \cos \alpha (\cos^2 \beta + \sin^2 \beta) \\
&= \cos \alpha
\end{aligned}
$$

57. Let $\alpha = \sin^{-1}\left(\frac{3}{5}\right)$ or $\sin\alpha = \frac{3}{5}$ and $\beta = \cos^{-1}\left(\frac{4}{5}\right)$ or $\cos\alpha = \frac{4}{5}$. Then

$$
\begin{aligned}
\cos(\alpha + \beta) &= \cos\alpha\cos\beta - \sin\alpha\sin\beta \\
&= (\sqrt{1 - \sin^2\alpha})\frac{4}{5} - \frac{3}{5}(\sqrt{1 - \cos^2\beta}) \\
&= \left(\sqrt{1 - \frac{9}{25}}\right)\frac{4}{5} - \frac{3}{5}\left(\sqrt{1 - \frac{16}{25}}\right) \\
&= \frac{16}{25} - \frac{9}{25} = \frac{7}{25}
\end{aligned}
$$

59. Let $\alpha = \cos^{-1}\left(\frac{2}{3}\right)$ or $\cos\alpha = \frac{2}{3}$ and $\beta = \sin^{-1}\left(\frac{1}{2}\right)$ or $\sin\alpha = \frac{1}{2}$. Then

$$
\begin{aligned}
\sin(\alpha + \beta) &= \sin\alpha\cos\beta + \cos\alpha\sin\beta \\
&= (\sqrt{1 - \cos^2\alpha})(\sqrt{1 - \sin^2\beta}) + \left(\frac{2}{3}\right)\left(\frac{1}{2}\right) \\
&= \frac{\sqrt{15}}{6} - \frac{2}{6} = \frac{\sqrt{15} + 2}{6}
\end{aligned}
$$

61. Given a triangle with angles α, β, and γ, then $\alpha + \beta + \gamma = 180°$ and

$$
\begin{aligned}
\tan(\alpha + \beta) &= \tan(180° - \tan\gamma) \\
&= \frac{\tan 180° - \tan\gamma}{1 + \tan 180° \tan\gamma} \\
&= \frac{0 - \tan\gamma}{1 + 0} \\
&= -\tan\gamma
\end{aligned}
$$

63. Given a triangle with angles α, β, and γ, then $\alpha + \beta + \gamma = 180°$ and

$$
\begin{aligned}
\sin\left(\frac{\alpha + \beta}{2}\right) &= \cos\left(90° - \frac{\alpha + \beta}{2}\right) \\
&= \cos\left(\frac{180° - \alpha - \beta}{2}\right) \\
&= \cos\frac{\gamma}{2}
\end{aligned}
$$

65. Label α, β, and γ at angles A, B, and C, respectively. Since $\triangle ABC$ is isoceles, $\alpha = \beta$ and from the given figure $\gamma = 150°$; since $\alpha + \beta + \gamma = 180°$, we have $\alpha = \beta = 15°$ so

$$\cos 15° = \frac{|AD|}{|AB|} = \frac{|AC| + |CD|}{|AB|}$$

Now, $|AC| = 2$; $\sin 30° = \frac{|BD|}{2}$, so $|BD| = 1$; $\cos 30° = \frac{|CD|}{2}$, so $|CD| = \sqrt{3}$; Finally,

$$|AB|^2 = (|AC| + |CD|)^2 + |BD|^2 = (2 + \sqrt{3})^2 + 1 = 8 + 4\sqrt{3}$$

so that $|AB| = \sqrt{8 + 4\sqrt{3}} = 2\sqrt{2 + \sqrt{3}}$. Substituting these values

$$\cos 15° = \frac{|AC| + |CD|}{|AB|} = \frac{2 + \sqrt{3}}{2\sqrt{2 + \sqrt{3}}} = \frac{2 + \sqrt{3}}{2\sqrt{2 + \sqrt{3}}} \cdot \frac{\sqrt{2 + \sqrt{3}}}{\sqrt{2 + \sqrt{3}}} = \frac{\sqrt{2 + \sqrt{3}}}{2}$$

67.

$$\frac{\cos(x + h) - \cos x}{h} = \frac{\cos x \cos h - \sin x \sin h - \cos x}{h}$$

$$= -\sin x \left(\frac{\sin h}{h}\right) - \cos x \left(\frac{1 - \cos h}{h}\right)$$

Problem Set 4.4

1. $\cos 2\theta = \cos^2 \theta - \sin^2 \theta$, not $\sin^2 \theta + \cos^2 \theta$.

3. $\cos \frac{1}{2}\theta = \pm\sqrt{\frac{1 + \cos \theta}{2}}$, not $\frac{\cos \theta}{2}$.

5.

$$\tan(45° + \theta) = \frac{\frac{\sqrt{2}}{2} + \tan \theta}{1 - \frac{\sqrt{2}}{2} \tan \theta}$$

7. $\tan 2\theta = \tan(\theta + \theta)$, not $\tan \theta + \tan \theta$.

9.

$$\tan \frac{1}{2}\theta = \frac{\sin \theta}{1 + \cos \theta},$$

$$= \frac{\sin \theta}{1 + \cos \theta} \cdot \frac{1 - \cos \theta}{1 - \cos \theta}$$

$$= \frac{\sin \theta(1 - \cos \theta)}{1 - \cos^2 \theta}$$

$$= \frac{\sin \theta(1 - \cos \theta)}{\sin^2 \theta}$$

$$= \frac{1 - \cos \theta}{\sin \theta}$$

11.

$$\frac{2\tan\left(\frac{1}{2}\theta\right)}{1 - \tan^2\left(\frac{1}{2}\theta\right)} = \tan\left(\frac{1}{2} \cdot 2\theta\right) = \tan \theta \neq \frac{2}{1 + \tan \theta}$$

13. a.   quadrant III or IV

     b.   quadrant I

15. a.   quadrant I or IV

     b.   quadrant I or II

17. quadrant II

19. a.   Use $2\cos^2\theta - 1 = \cos 2\theta$ and if $\theta = 22.5°$ then $2\cos^2(22.5°) - 1 = \cos 2(22.5) = \cos 45° = \frac{\sqrt{2}}{2}$.

     b.   Use $\tan \frac{1}{2}\theta = \frac{1 - \cos \theta}{\sin \theta}$ and if $\theta = 45°$ then $\tan(22.5°) = \frac{1 - \cos 45°}{\sin 45°} = \frac{1 - \sqrt{2}/2}{\sqrt{2}/2} = \sqrt{2} - 1$.

21. a.   Use $\tan 2\theta = \frac{2\tan \theta}{1 - \tan^2 \theta}$ and if $\theta = \frac{\pi}{8}$ then $\frac{2\tan \frac{\pi}{8}}{1 - \tan^2 \frac{\pi}{8}} = \tan 2\left(\frac{\pi}{8}\right) = \tan \frac{\pi}{4} = 1$.

     b.   Use $\sin \frac{1}{2}\theta = \sqrt{\frac{1 - \cos \theta}{2}}$ and if $\theta = 60°$ then $\sqrt{\frac{1 - \cos 60°}{2}} = \sin \frac{1}{2}(60°) = \sin 30° = \frac{1}{2}$.

23. $\cos \theta = \frac{\sqrt{5}}{5}$; $\sin \theta = \frac{2\sqrt{5}}{5}$; $\tan \theta = 2$

25. $\cos \theta = \frac{\sqrt{10}}{10}$; $\sin \theta = \frac{3\sqrt{10}}{10}$; $\tan \theta = 3$

27. $\cos \theta = \frac{1}{2}$; $\sin \theta = \frac{\sqrt{3}}{2}$; $\tan \theta = \sqrt{3}$

29. $\cos 2\theta = 2(1 - \sin^2 \theta) - 1$ and using $\sin \theta = \frac{3}{5}$ we get $1 - \frac{18}{25} = \frac{7}{25}$; $\sin 2\theta = 2\sin \theta\sqrt{1 - \sin^2 \theta}$ and using $\sin \theta = \frac{3}{5}$ we get $\sin 2\theta = \frac{24}{25}$; $\tan 2\theta = \frac{\sin 2\theta}{\cos 2\theta} = \frac{24}{7}$

31. Using $\tan = -\frac{5}{12}$ and $\tan 2\theta = \dfrac{2\tan\theta}{1-\tan^2\theta} = -\frac{120}{119}$ and since $\tan 2\theta = \dfrac{\sin 2\theta}{\cos 2\theta} = \dfrac{-120/x}{119/x}$ and $\sin^2 2\theta + \cos^2 2\theta =$ 1 we get $\left(\frac{-120}{x}\right)^2 + \left(\frac{119}{x}\right)^2 = 1$ and solving for x gives $x = 169$ so $\cos 2\theta = \frac{119}{169}$ and $\sin 2\theta = \frac{-120}{169}$.

33. $\cos 2\theta = 2\left(\frac{5}{9}\right)^2 - 1 = \frac{-31}{81}$; $\sin 2\theta = 2\cos\theta\sqrt{1-\cos^2\theta} = \frac{20\sqrt{14}}{81}$; $\tan 2\theta = -\frac{20\sqrt{14}}{31}$

35. $\cos\frac{1}{2}\theta = \sqrt{\dfrac{1-\sqrt{1-\sin^2\theta}}{2}}$ and substituting $\sin\theta = \frac{3}{5}$ gives $\cos\frac{1}{2}\theta = \frac{1}{\sqrt{10}}$; $\sin\frac{1}{2}\theta = \sqrt{\dfrac{1+\sqrt{1-\sin^2\theta}}{2}}$ and substituting $\sin\theta = \frac{3}{5}$ gives $\sin\frac{1}{2}\theta = \frac{3}{\sqrt{10}}$; $\tan\frac{1}{2}\theta = \frac{1}{3}$

37. $\sin\frac{1}{2}\theta = \sqrt{\frac{1+12/13}{2}} = -\frac{5}{\sqrt{26}}$; $\cos\frac{1}{2}\theta = \sqrt{\frac{1-12/13}{2}} = \frac{1}{\sqrt{26}}$; $\tan\frac{1}{2}\theta = 5$

39. $\cos\frac{1}{2}\theta = \frac{2}{3\sqrt{2}}$; $\sin\frac{1}{2}\theta = \frac{\sqrt{14}}{3\sqrt{2}}$; $\tan\frac{1}{2}\theta = \frac{2}{\sqrt{14}}$

41.

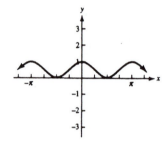

43.

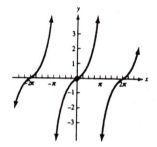

45.

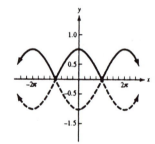

47. a. $\sin\frac{1}{2}\left(\frac{\pi}{6}\right) = M^{-1}$, so $M \approx 3.9$.

   b. $\sin\left(\frac{\pi}{12}\right) = \sin(15°)$ and $\sin(45° + 15°) = \sqrt{6} + \sqrt{2}$.

   c. From the equation given $\sin\frac{1}{2}\theta = M^{-1}$ take the inverse of the sine to get $\frac{1}{2}\theta = \sin^{-1}\left(\frac{1}{M}\right)$ then $\theta = 2\sin^{-1}\left(\frac{1}{M}\right)$.

49. Since $x = (v_0 \cos \theta)t$ and $y = (v_0 \sin \theta)t - 16t^2$ and the maximum range would be when $y = 0$ so solving for t from $0 = (v_0 \sin \theta)t - 16t^2$ gives $t = \dfrac{v_0 \sin \theta}{16}$ and substituting t into the equation for x gives $x = \frac{1}{32}v_0^2 \sin 2\theta$. Since $v_0 = 256$, $x \approx 2048$ ft when the angle of elevation is $45°$ because the max is greatest when $\sin 2\theta = 1$ or $2\theta = 90°$ or $\theta = 45°$.

51.
$$2 \sin \frac{\alpha}{2} \cos \frac{\alpha}{2} = \sin \left(\frac{\alpha}{2} + \frac{\alpha}{2} \right)$$
$$= \sin \alpha$$

53.
$$\frac{2 \tan \theta}{1 + \tan^2 \theta} = \frac{2 \frac{\sin \theta}{\cos \theta}}{1 + \frac{\sin^2 \theta}{\cos^2 \theta}}$$
$$= \frac{\frac{2 \sin \theta}{\cos \theta}}{\frac{\cos^2 + \sin^2 \theta}{\cos^2 \theta}}$$
$$= \frac{2 \sin \theta}{\cos \theta} \cdot \frac{\cos^2 \theta}{1}$$
$$= \sin 2\theta$$

55.
$$\tan \frac{1}{2}\theta = \frac{\sin \frac{1}{2}\theta}{\cos \frac{1}{2}\theta}$$
$$= \frac{\pm \sqrt{\frac{1 - \cos \theta}{2}}}{\pm \sqrt{\frac{1 + \cos \theta}{2}}}$$
$$= \pm \sqrt{\frac{1 - \cos \theta}{2} \cdot \frac{2}{1 + \cos \theta} \cdot \frac{1 - \cos \theta}{1 - \cos \theta}}$$
$$= \pm \sqrt{\frac{(1 - \cos)^2}{1 - \cos^2 \theta}}$$
$$= \pm \frac{\sqrt{(1 - \cos \theta)^2}}{\sqrt{\sin^2 \theta}}$$
$$= \frac{1 - \cos \theta}{\sin \theta}$$

57.

$$\sin 2\theta \sec \theta = 2 \sin \theta \cos \theta \cdot \frac{1}{\cos \theta}$$
$$= 2 \sin \theta$$

59. From Figure 4.17, $\sin 2\theta = \dfrac{|CD|}{|CO|} = \dfrac{|CD|}{1} = |CD|$. Then using the right triangle ADC, $\sin \theta = \dfrac{|CD|}{2 \cos \theta}$. Thus, $2 \sin \theta \cos \theta = |CD|$. So, $\sin 2\theta = 2 \sin \theta \cos \theta$.

61. **a.** $|AO| = |BO| = r$, so $\alpha = \beta$ (base angles in an isoceles triangle.) Also, $\alpha + \beta + \gamma = 180°$, and $\gamma = 180° - \theta$, so

$$\alpha + \alpha + (180° - \theta) = 180°$$
$$2\alpha - \theta = 0$$
$$\alpha = \frac{1}{2}\theta$$

b. $\sin \theta = \dfrac{|BD|}{|BO|}$; $\cos \theta = \dfrac{|OD|}{|BO|}$;

$$\tan \alpha = \frac{|BD|}{|AD|} = \frac{|BD|}{|AO| + |OD|} = \frac{|BD|}{r + |BO| \cos \theta}$$
$$= \frac{|BO| \sin \theta}{r + |BO| \cos \theta} = \frac{r \sin \theta}{r + r \cos \theta} = \frac{\sin \theta}{1 + \cos \theta}$$

Thus, $\tan \frac{1}{2}\theta = \dfrac{\sin \theta}{1 + \cos \theta}$.

Problem Set 4.5

1. $\cos(35° - 24°) - \cos(35° + 24°) = \cos(11°) - \cos(59°)$

3. $\sin(70° + 24°) - \sin(70° - 24°) = \sin(94°) - \sin(46°)$

5. $\cos(2\theta - 4\theta) - \cos(2\theta + 4\theta) = \frac{1}{2} \cos(-2\theta) - \frac{1}{2} \cos 6\theta$

7. $\frac{1}{2}(2 \cos \theta \cos 3\theta) = \frac{1}{2}(\cos(\theta - 3\theta) + \cos(4\theta)) = \frac{1}{2} \cos +2\theta + \frac{1}{2} \cos 4\theta$

9. Use $\sin x + \sin y = 2\sin\left(\frac{x+y}{2}\right)\cos\left(\frac{x-y}{2}\right)$. Then

$$\sin 43° + \sin 64° = 2\sin\left(\frac{43° + 64°}{2}\right)\cos\left(\frac{43° - 64°}{2}\right)$$
$$= 2\sin 53.5\cos(-10.5)$$
$$= 2\sin 53.5\cos 10.5$$

11. Use $\sin x + \sin y = 2\sin\left(\frac{x+y}{2}\right)\cos\left(\frac{x-y}{2}\right)$. Then

$$\sin 15° + \sin 30° = 2\sin\left(\frac{15° + 30°}{2}\right)\cos\left(\frac{15° - 30°}{2}\right)$$
$$= 2\sin 22.5\cos(-7.5)$$
$$= 2\sin 22.5\cos 7.5$$

13. Use $\cos x + \cos y = 2\cos\left(\frac{x+y}{2}\right)\cos\left(\frac{x-y}{2}\right)$. Then

$$\cos 6x + \cos 2x = 2\cos\left(\frac{6x + 2x}{2}\right)\cos\left(\frac{6x - 2x}{2}\right)$$
$$= 2\cos 4x\cos 2x$$

15. Use $\cos x + \cos y = 2\cos\left(\frac{x+y}{2}\right)\cos\left(\frac{x-y}{2}\right)$. Then

$$\cos 5y + \cos 9y = 2\cos\left(\frac{5y + 9y}{2}\right)\cos\left(\frac{5y - 9y}{2}\right)$$
$$= 2\cos 7y\cos(-2y)$$
$$= 2\cos 7y\cos 2y$$

17. $a = 1$ and $b = 1$, so $\sqrt{a^2 + b^2} = \sqrt{2}$. $\cos\alpha = \frac{1}{\sqrt{2}}$ and $\sin\alpha = \frac{1}{\sqrt{2}}$, then $\alpha = 45°$ and $\sin\theta + \cos\theta = \sqrt{2}\sin(\theta + 45°)$.

19. $a = -1$ and $b = 1$, so $\sqrt{a^2 + b^2} = \sqrt{2}$. $\cos\alpha = -\frac{1}{\sqrt{2}}$ and $\sin\alpha = \frac{1}{\sqrt{2}}$, then since the coordinates are in quadrant II, $\alpha = \frac{3\pi}{4}$ and $-\sin\frac{\theta}{2} + \cos\frac{\theta}{2} = \sqrt{2}\sin\left(\frac{\theta}{2} + \frac{3\pi}{4}\right)$.

21. $a = -1$ and $b = \sqrt{3}$, so $\sqrt{a^2 + b^2} = \sqrt{4} = 2$. $\cos\alpha = -\frac{1}{2}$ and $\sin\alpha = \frac{\sqrt{3}}{2}$, then $\alpha = \frac{2\pi}{3}$ and $-\sin\pi\theta + \sqrt{3}\cos\pi\theta = 2\sin\left(\pi\theta + \frac{2\pi}{3}\right)$.

23.

$$\frac{\sin\theta + \sin 3\theta}{2\sin 2\theta} = 2\sin\left(\frac{\theta + 3\theta}{2}\right)\cos\left(\frac{\theta - 3\theta}{2}\right)$$

$$= \frac{2\sin 2\theta \cos(-\theta)}{2\sin 2\theta}$$

$$= \cos(-\theta)$$

$$= \cos\theta$$

25.

$$\frac{\sin\theta + \sin 3\theta}{4\cos^2\theta} = \frac{\sin 2\theta}{2\cos\theta}$$

$$= \frac{\sin(\theta + \theta)}{2\cos\theta}$$

$$= \frac{\sin\theta\cos\theta + \sin\theta\cos\theta}{2\cos\theta}$$

$$= \sin\theta$$

27.

$$\frac{\sin 5\theta + \sin 3\theta}{\cos 5\theta + \cos 3\theta} = \frac{2\sin\left(\frac{5\theta + 3\theta}{2}\right)\cos\left(\frac{5\theta - 3\theta}{2}\right)}{2\cos\left(\frac{5\theta + 3\theta}{2}\right)\cos\left(\frac{5\theta - 3\theta}{2}\right)}$$

$$= \frac{\sin 4\theta}{\cos 4\theta}$$

$$= \tan 4\theta$$

29.

$$\frac{\cos 5\omega + \cos\omega}{\cos\omega - \cos 5\omega} = \frac{2\cos\left(\frac{5\omega + \omega}{2}\right)\cos\left(\frac{5\omega - \omega}{2}\right)}{-2\cos\left(\frac{5\omega + \omega}{2}\right)\cos\left(\frac{5\omega - \omega}{2}\right)}$$

$$= \frac{2\cos 3\omega \cos 2\omega}{2\sin 3\omega \sin 2\omega}$$

$$= \frac{\cot 2\omega}{\tan 3\omega}$$

31.

$$\frac{\sin x + \sin y}{\cos x + \cos y} = \frac{2 \sin \left(\frac{x+y}{2}\right) \cos \left(\frac{x-y}{2}\right)}{2 \cos \left(\frac{x+y}{2}\right) \cos \left(\frac{x-y}{2}\right)}$$

$$= \frac{2 \sin \left(\frac{x+y}{2}\right)}{2 \cos \left(\frac{x+y}{2}\right)}$$

$$= \tan \left(\frac{x+y}{2}\right)$$

33.

$$\cos^2 \frac{\theta}{2} - \sin^2 \frac{\theta}{2} = \frac{1 + \cos \theta}{2} - \frac{1 - \cos \theta}{2}$$

$$= \frac{2 \cos \theta}{2}$$

$$= \cos \theta$$

35. $a = -1$, $b = 1$, $\sqrt{a^2 + b^2} = \sqrt{2}$, $\cos \alpha = -\frac{1}{\sqrt{2}}$, $\sin \alpha = \frac{1}{\sqrt{2}}$, and $\alpha = \frac{3\pi}{4}$. So, $y = \sqrt{2} \sin \left(\theta + \frac{3\pi}{4}\right)$. The starting point is $\left(-\frac{3\pi}{4}, 0\right)$, amplitude $\sqrt{2}$, and period 2π.

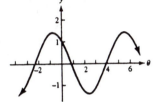

37. $a = 2$, $b = -3$, $\sqrt{a^2 + b^2} = \sqrt{13}$, $\cos \alpha = \frac{2}{\sqrt{13}}$, $\sin \alpha = -\frac{3}{\sqrt{13}}$, and $\alpha = \frac{38\pi}{45} \approx 5.3$. So, $y = \sqrt{13} \sin(\theta + 5.3)$. The starting point is $(-5.3, 0)$, amplitude $\sqrt{13}$, $b = 1$, and period 2π.

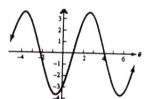

39. $a = 8$, $b = 15$, $\sqrt{a^2 + b^2} = 17$, $\cos \alpha = \frac{8}{17}$, $\sin \alpha = \frac{15}{17}$, and $\alpha = 1.1$. So, $y = 17 \sin(\theta + 1.1)$. The starting point is $(-1.1, 0)$, amplitude 17, $b = 1$, and period 2π.

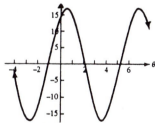

41. $2 \sin \left(\frac{x-3x}{2}\right) \cos \left(\frac{x+3x}{2}\right) = 2 \sin(-x) \cos 2x = (-2 \sin x)(\cos 2x) = 0$. When $2 \sin x = 0$, then $x = 0, \pi$ and when $\cos 2x = 0$, then $x = \frac{\pi}{4}, \frac{3\pi}{4}, \frac{5\pi}{4}, \frac{7\pi}{4}$.

43. $-2 \sin \left(\frac{5\alpha + 3\alpha}{2}\right) \sin \left(\frac{5\alpha + 3\alpha}{2}\right) = 2 \sin 4\alpha \sin \alpha = 0$. When $\sin 4\alpha = 0$, then $\alpha = 0, \frac{\pi}{4}, \frac{\pi}{2}, \frac{3\pi}{4}, \pi, \frac{5\pi}{4}, \frac{3\pi}{2}, \frac{7\pi}{4}$ and when $\sin \alpha = 0$, then $\alpha = 0, \pi$.

45. $1 - \sin^2 x - 3\sin x + 3 = -\sin^2 x - 3\sin x + 4 = (-\sin x + 1)(\sin x + 4) = 0$. When $-\sin x = -1$, then $x = \frac{\pi}{2}$.
Since $\sin x$ cannot be -4 then there is only one x.

47.
$$2\sin\alpha\cos\beta = 2\cos\beta\sin\alpha$$
$$= \sin(\beta + \alpha) - \sin(\beta - \alpha)$$
$$= \sin(\alpha + \beta) + \sin(\alpha - \beta)$$

49. Let $x = \alpha + \beta$ and $y = \alpha - \beta$ then $2\sin\alpha\cos\beta = \sin(\alpha + \beta) + \sin(\alpha - \beta) = \sin x + \sin y$ and combining x and y and solving for α gives $\alpha = \frac{x+y}{2}$ and $\beta = \frac{x-y}{-2}$ and then substituting α and β into the $2\sin\alpha\cos\beta$ gives $\sin x + \sin y$ which is identity 35.

51. $y = 3\sin x + 3\cos x = 3\sqrt{2}\sin(x + \alpha)$ and the amplitude is $3\sqrt{2}$ and the period is 2π.

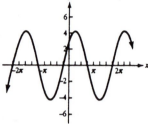

53. $y = \sqrt{2}\left(\sin\frac{x}{2} + \cos\frac{x}{2}\right) = 2\sin\frac{1}{2}\left(x + \frac{1}{2}\alpha\right)$ and the amplitude is 2 and the period is 4π.

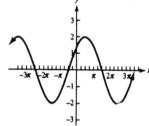

55. $y = 2\sin(1906\pi x)\cos(512\pi x)$

57.   a.

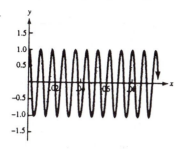

b.

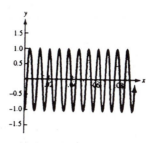

c.

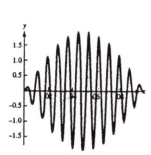

d. $y = -2\sin(234\pi x)\sin(10\pi x)$

59. $y = \cot 4x$

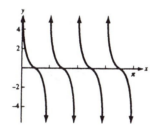

Chapter 4 Sample Test

1.

$$
\begin{aligned}
\frac{\sec^2\theta + \tan^2\theta + 1}{\sec\theta} &= \frac{\sec^2\theta + \sec^2\theta}{\sec\theta} \\
&= \frac{2\sec^2\theta}{\sec\theta} \\
&= 2\sec\theta
\end{aligned}
$$

3.
$$\frac{2\tan\frac{\pi}{6}}{1-\tan^2\frac{\pi}{6}} = \tan\frac{\pi}{6} \approx \sqrt{3}$$

5.
$$\frac{\csc^2\alpha}{1+\cot^2\alpha} = \frac{\cot^2\alpha+1}{1+\cot^2\alpha} = 1$$

7.
$$\frac{1}{\sin\gamma+\cos\gamma} + \frac{1}{\sin\gamma-\cos\gamma} = \frac{\sin\gamma-\cos\gamma+\sin\gamma+\cos\gamma}{\sin^2\gamma-\cos^2\gamma}$$
$$= \frac{2\sin\gamma}{\sin^2\gamma-\cos^2\gamma} \cdot \frac{\sin^2\gamma+\cos^2\gamma}{\sin^2\gamma+\cos^2\gamma}$$
$$= \frac{2\sin\gamma(\sin^2\gamma+\cos^2\gamma)}{\sin^4\gamma-\cos^4\gamma}$$
$$= \frac{2\sin\gamma}{\sin^4\gamma-\cos^4\gamma}$$

9.
$$\frac{\sin 5\theta+\sin 3\theta}{\cos 5\theta-\cos 3\theta} = \frac{2\sin 4\theta\cos\theta}{-2\sin 4\theta\sin\theta} = -\cot\theta$$

11. Let $t = \frac{\pi}{4}$, then $\left(\sin\frac{\pi}{4} + \cos\frac{\pi}{4}\right) = 2$ and not 1.

13. $\cos^2\theta = \frac{1}{4}$ and $\cos\theta = \frac{1}{\sqrt{4}}$, so $\theta = \frac{\pi}{3}, \frac{2\pi}{3}, \frac{4\pi}{3}, \frac{5\pi}{3}$.

15. $\sin^2\theta + 3\sin\theta - 1 = 0$ and using the quadratic equation and then solving for θ gives $\theta = 0.31$ and $\theta = 2.83$.

17. 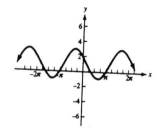   **19.**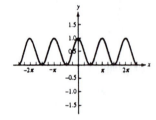

21. $\cot 2\theta = \frac{1-\tan^2\theta}{2\tan\theta} = -\frac{4}{3}$ and simplifying gives $\tan^2\theta - \frac{8}{3}\tan\theta - 1 = 0$. Solving for $\tan\theta$ gives $\tan\theta = 3$ or $-\frac{1}{2}$ but since this is in quadrant I $\tan\theta = 3$.

23. $|P_\alpha P_\beta| = \sqrt{(\cos\beta-\cos\alpha)^2 + (\sin\beta-\sin\alpha)^2} = \sqrt{2 - 2(\cos\alpha\cos\beta + \sin\alpha\sin\beta)}$

25. Let P_α and P_β be defined as in Problem 23. Then, $|P_\alpha P_\beta| = \sqrt{2 - 2(\cos\alpha\cos\beta + \sin\alpha\sin\beta)}$. The angle between rays through P_α and P_β is $\alpha - \beta$, so $|P_\alpha P_\beta| = \sqrt{2 - 2\cos(\alpha - \beta)}$ from Problem 23. Thus,

$$\sqrt{2 - 2\cos(\alpha - \beta)} = \sqrt{2 - 2(\cos\alpha\cos\beta + \sin\alpha\sin\beta)}$$
$$2 - 2\cos(\alpha - \beta) = 2 - 2(\cos\alpha\cos\beta + \sin\alpha\sin\beta)$$
$$\cos(\alpha - \beta) = \cos\alpha\cos\beta + \sin\alpha\sin\beta$$

27. $\sin 2\theta = 2\sin\theta\cos\theta = 2\sqrt{1 - \cos^2\theta}\cos\theta$ substituting in $\cos\theta = -\frac{4}{5}$ gives $\sin 2\theta = -\frac{24}{25}$.

29. Since $\sin\alpha\cos\beta = \frac{1}{2}[\sin(\alpha + \beta) - \sin(\alpha - \beta)]$, we have $\alpha = 3$ and $\beta = 1$ so $\frac{1}{2}[\sin 4\theta + \sin 2\theta]$.

Chapter 4 Miscellaneous Problems

1.

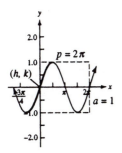

3.

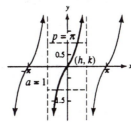

5.

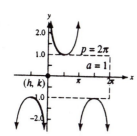

7. $y = \sin\frac{\pi x}{20}$

9. $y = \sin\frac{\pi x}{20} + \sin\left(\frac{\pi x}{20} - \frac{\pi}{2}\right)$

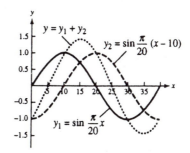

11. $\cos 2\theta = 1 - 2\sin^2 \theta = 1 - 2\left(\frac{4}{5}\right)^2 = -\frac{7}{25}$

13. a. $\cos(30°) = \frac{\sqrt{3}}{2}$ b. $\sin(45°) = \frac{\sqrt{2}}{2}$ c. $\tan(-60°) = -\sqrt{3}$

15. a. $\tan -2(15°) = -\frac{\sqrt{3}}{3}$

   b. $\sin 10(15°) = \frac{1}{2}$

   c. $[\cos 4(15°)]^2 = \left(\frac{1}{2}\right)^2 = \frac{1}{4}$

17. a. $\cos 2\theta = 2\cos^2 \theta - 1$ and $\cos \theta = \frac{4}{5}$, so $2\left(\frac{4}{5}\right)^2 - 1 = \frac{9}{25}$.

   b. $\sin 2\theta = 2\cos \theta \sin \theta$ and $\cos \theta = \frac{4}{5}$ and $\sin \theta = \frac{3}{5}$, so $2\left(\frac{4}{5}\right)\left(\frac{3}{5}\right) = \frac{24}{25}$.

   c. From parts (a) and (b), $\tan 2\theta = \frac{\sin 2\theta}{\cos 2\theta} = \frac{24}{9}$.

19. a. $\tan 2\theta = \frac{24}{9}$ from Problem 17(c), so $\cot 2\theta = \frac{9}{24}$

   b. $\cos 2\theta = \frac{9}{25}$ from Problem 17(a), so $\sec 2\theta = \frac{25}{9}$

   c. $\sin 2\theta = \frac{24}{25}$ from Problem 17(b), so $\sec 2\theta = \frac{25}{24}$

21. After canceling the $\sin^2 \theta$ from both sides, this gives $-\cos \theta = 1$ and $\theta = \pi$.

23. After canceling a $\tan 2\theta$ from both sides, this gives $2\theta \approx 1.249$ and $\theta = 0.62$.

25. After canceling $3\sin 4x$ and $-\sin 3x$ from both sides, this gives $\sqrt{3} = 2\sin 2x$ and $x = \frac{\pi}{6}$.

27.

$$
\begin{aligned}
\frac{\sec u}{\tan^2 u} &= \frac{1/\cos u}{\sin^2 u / \cos^2 u} \\
&= \frac{1}{\cos u} \cdot \frac{\cos^2 u}{\sin^2 u} \\
&= \frac{\cos u}{\sin^2 u} \\
&= \frac{\cos u}{\sin u} \cdot \frac{1}{\sin u} \\
&= \cot u \csc u
\end{aligned}
$$

29. $\sec(-\theta) = \dfrac{1}{\cos(-\theta)} = \dfrac{1}{\cos \theta} = \sec \theta$

31.

$$\frac{\sin \beta}{\cos \beta} = \frac{\sin \beta}{\cos \beta} + \frac{\cos \beta}{\sin \beta}$$

$$= \frac{\sin^2 \beta + \cos^2 \beta}{\cos \beta \sin \beta}$$

$$= \frac{1}{\cos \beta \sin \beta}$$

$$= \sec \beta \csc \beta$$

33.

$$2 \sin \left(\frac{1}{2}\alpha\right) \cos \left(\frac{1}{2}\alpha\right) = \sin \left(\frac{1}{2}\alpha + \frac{1}{2}\alpha\right) = \sin \alpha$$

35.

$$\sin 2\theta \sec \theta = 2 \sin \theta \cos \theta \cdot \frac{1}{\cos \theta} = 2 \sin \theta$$

37.

$$\frac{\csc \gamma + 1}{\csc \gamma \cos \gamma} = \frac{\csc \gamma}{\csc \gamma \cos \gamma} + \frac{1}{\csc \gamma \cos \gamma}$$

$$= \frac{1}{\cos \gamma} + \frac{1}{\cos \gamma / \sin \gamma}$$

$$= \sec \gamma + \frac{\sin \gamma}{\cos \gamma}$$

$$= \sec \gamma + \tan \gamma$$

39.

$$\frac{\cot 3\theta - \sin 3\theta}{\sin 3\theta} = \frac{(\cot 3\theta - \sin 3\theta) \sin 3\theta}{\sin^2 3\theta}$$

$$= \frac{\frac{\cos 3\theta}{\sin 3\theta} \cdot \sin 3\theta - \sin^2 3\theta}{\sin^2 3\theta}$$

$$= \frac{\cos 3\theta - (1 - \cos^2 3\theta)}{\sin^2 3\theta}$$

$$= \frac{\cos^2 3\theta + \cos 3\theta - 1}{\sin^2 3\theta}$$

41.

$$(\sin \alpha \cos \alpha \cos \beta + \sin \beta \cos \beta \cos \alpha) \sec \alpha \sec \beta = \frac{\sin \alpha \cos \alpha \cos \beta + \sin \beta \cos \beta \cos \alpha}{\cos \alpha \cos \beta}$$

$$= \sin \alpha + \sin \beta$$

43.
$$\tan \alpha = \tan(\alpha - (\alpha - \beta)]$$
$$= \frac{\tan \alpha - \tan(\alpha - \beta)}{1 + \tan \alpha \tan(\alpha - \beta)}$$

45.
$$\cos 345° = \cos(315° + 30°) = \cos 315° \cos 30° - \sin 315° \sin 30°$$
$$= \left(\frac{\sqrt{2}}{2}\right)\frac{\sqrt{3}}{2} - \left(-\frac{\sqrt{2}}{2}\right)\frac{1}{2}$$
$$= \frac{\sqrt{6} + \sqrt{2}}{4}$$

47.
$$\sqrt{\frac{1 - \cos 270°}{2}} = \sin \frac{1}{2}(270°) = \sin 1135° = \frac{\sqrt{2}}{2}$$

49. Since $u = 6 \sin \theta$, then
$$\frac{1}{\sqrt{36 - 36 \sin^2 \theta}} = \frac{1}{6\sqrt{1 - \sin^2 \theta}} = \frac{1}{6\sqrt{\cos^2 \theta}} = \frac{1}{6} \sec \theta$$

51. Since $u = 10 \tan \theta$, then
$$\sqrt{100 \tan^2 + 100} = 10\sqrt{\tan^2 \theta + 1} = 10\sqrt{\sec^2 \theta} = 10 \sec \theta$$

53. Since $u = 8 \sec \theta$, then
$$\sqrt{64 \sec^2 \theta - 64} = 8\sqrt{\sec^2 \theta - 1} = 8\sqrt{\tan^2 \theta} = 8 \tan \theta$$

55. The point of intersection is $(\cos \alpha, \sin \alpha)$ and then $(\cos 30°, \sin 30°) = \left(\frac{\sqrt{3}}{2}, \frac{1}{2}\right)$.

57. Maximum values are at $x = 0, \pi, 2\pi$ and minimum values are at $x = \frac{x}{2}, \frac{3\pi}{2}$.

59. $\phi + \alpha + (180° - \beta) = 180°$ so $\phi = \beta - \alpha$ and

$$
\begin{aligned}
\tan \phi &= \tan(\beta - \alpha) \\
&= \frac{\tan \beta - \tan \alpha}{1 + \tan \beta \tan \alpha} \\
&= \frac{m_2 - m_1}{1 + m_1 m_2}
\end{aligned}
$$

CHAPTER 5   OBLIQUE TRIANGLES AND VECTORS

Problem Set 5.1

1.

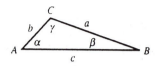

5.

$$\cos \alpha = \frac{b^2 + c^2 - a^2}{2bc}$$

$$= \frac{8^2 + 2^2 - 7^2}{(2)(8)(2)} = 0.59375$$

$$\alpha = 54°$$

7.

$$\cos \gamma = \frac{a^2 + b^2 - c^2}{2ab}$$

$$= \frac{10^2 + 4^2 - 8^2}{(2)(10)(4.0)} = 0.65$$

$$\gamma = 49°$$

9.

$$\cos \beta = \frac{a^2 + c^2 - b^2}{2ac}$$

$$= \frac{12^2 + 15^2 - 6^2}{(2)(12)(15)} = 0.925$$

$$\beta = 22°$$

11.

$$c^2 = a^2 + b^2 - 2ab\cos\gamma$$
$$= 18^2 + 25^2 - 2(18)(25)\cos 30° = 169.577$$
$$c \approx 13$$

13.

$$c^2 = a^2 + b^2 - 2ab\cos\gamma$$
$$= 15^2 + 8^2 - 2(15)(8)\cos 38° = 99.877$$
$$c \approx 10$$

15.

$$a^2 = b^2 + c^2 - 2bc\cos\alpha$$
$$= 14^2 + 12^2 - 2(14)(12)\cos 82° = 293.238$$
$$a \approx 17$$

17. Since $b^2 = a^2 + c^2 - 2ac\cos\beta$, substituting in the known values gives $(16.3)^2 = (14.2)^2 + c^2 - 2(14.2)c\cos 115°$ and $265.69 = 201.64 + c^2 - 28.4c\cos 115°$. Then solving for c gives

$$c = \frac{28.4\cos 115° \pm \sqrt{(-28.4\cos 115°)^2 - 4(1)(-64.05)}}{2(1)}$$

This gives $c = 4.00$ or $c = -16$, and we know that distance is always positive, so $c = 4$ is correct. Next we can solve for α,

$$\cos\alpha = \frac{b^2 + c^2 - a^2}{2bc}$$
$$= \frac{(16.3)^2 + 4^2 - (14.2)^2}{2(16.3)(4)} = 0.61388$$

So, $\alpha = 52.1°$. Then solving for γ,

$$\gamma = 180° - \alpha - \beta$$
$$= 180° - 52.1° - 115°$$

So $\gamma = 12.9°$.

19. Since $a^2 = b^2 + c^2 - 2bc \cos \alpha$ substituting in the known values gives $(5.0)^2 = (4.0)^2 + c^2 - 2(4.0)c \cos 125°$ and $c^2 - 8.0c \cos 125° - 9 = 0$. Solving for c gives

$$c = \frac{8.0 \cos 125° \pm \sqrt{(-8.0 \cos 125°)^2 - 4(1)(-9)}}{2(1)}$$

This gives $c = 1.482444$ or $c = -6.0710555$. Since a distance cannot be negative, then $c = 1.5$. Now we can find β,

$$\cos \beta = \frac{a^2 + c^2 - b^2}{2ac}$$
$$= \frac{(5.0)^2 + (1.482444)^2 - (4.0)^2}{(2)(5.0)(1.482444)}$$

Thus, $\beta = 41°$. And then $\gamma = 180° - 125° - 41° = 14°$.

21. Since $c^2 = a^2 + b^2 - 2ab \cos \gamma$ then substituting in known values gives $(52.2)^2 = a^2 + (82.5)^2 - 2(82.5)a \cos 32.1°$ and $a^2 - (a)(165) \cos 32.1° + 4081.41 = 0$. Solving for a gives

$$a = \frac{165 \cos 32.1° \pm \sqrt{(-165 \cos 32.1°)^2 - 4(1)(4081.41)}}{2(1)}$$

Then $a = 98.222358$ or $a = 41.552759$. There are two positive values for a then two triangles. Now we find β and γ, using the first $a = 98.2$.

$$\cos \beta = \frac{a^2 + c^2 - b^2}{2ac}$$
$$= \frac{(98.222358)^2 + (52.2)^2 - (82.5)^2}{(2)(98.222358)(52.2)}$$

and $\beta = 57.1°$. Then $\alpha = 180° - 57.1° - 32.1° = 90.8°$. Now we can find β and γ, using the second $a = 41.6$.

$$\cos \beta = \frac{a^2 + c^2 - b^2}{2ac}$$
$$= \frac{(41.552759)^2 + (52.2)^2 - (82.5)^2}{(2)(41.552759)(52.2)}$$

and $\beta = 122.9°$. Then $\alpha = 180° - 122.9° - 32.1° = 25.0°$.

23. Since $a^2 = b^2 + c^2 - 2bc \cos \alpha$ and substituting in the known values gives $(10.2)^2 = (11.8)^2 + c^2 - 2(11.8)c \cos 47.0°$ and $c^2 - (c)(23.6) \cos 47.0° + 35.2 = 0$. Solving for c gives

$$c = \frac{23.6 \cos 47.0° \pm \sqrt{(-23.6 \cos 47.0°)^2 - 4(1)(35.2)}}{2(1)}$$

and $c = 13.484818$ or $c = 2.6103429$. There are two positive values for c and therefore, two triangles. Now we can find β and γ, using the first $c = 13.5$.

$$\cos \beta = \frac{a^2 + c^2 - b^2}{2ac}$$
$$= \frac{(13.484818)^2 + (10.2)^2 - (11.8)^2}{(2)(10.2)(13.484818)}$$

and $\beta = 57.8°$. Then $\alpha = 180° - 47.0° - 57.8° = 75.2°$. Now we can find β and γ, using the second $c = 2.61$.

$$\cos \beta = \frac{a^2 + c^2 - b^2}{2ac}$$
$$= \frac{(10.2)^2 + (2.6103429)^2 - (11.8)^2}{(2)(2.6103429)(10.2)}$$

and $\beta = 122.2°$. Then $\alpha = 180° - 122.2° - 47.0° = 10.8°$.

25. This problem cannot be solved, since the sum of the two sides must be greater than the hypotenuse, and this is not the case.

27. Since $a^2 = b^2 + c^2 - 2bc \cos \alpha$ and substituting in the known values gives $(5.2)^2 + (3.4)^2 - 2(5.2)(3.4) \cos 54.6°$ then $a = 4.26$. We now compute β and γ.

$$\cos \beta = \frac{a^2 + b^2 - c^2}{2ac}$$
$$= \frac{(4.26)^2 + (3.4)^2 - (5.2)^2}{2(4.26)(3.4)}$$

and $\beta = 84.7°$. Then $\gamma = 180° - \alpha - \beta = 180° - 54.6° - 40.7° = 41°$.

29. Since $c^2 = a^2 + b^2 - 2ab \cos \gamma$ substituting in the known values gives $(214)^2 + (320)^2 - 2(214)(320) \cos 14.8°$ so $c = 125.6$. We now compute β and α.

$$\cos \alpha = \frac{b^2 + c^2 - a^2}{2bc}$$
$$= \frac{(320)^2 + (125.6)^2 - (214)^2}{2(320)(125.6)}$$

and $\alpha = 84.7°$. Then $\beta = 180° - \alpha - \gamma = 180° - 25.8° - 14.8° = 139.4°$.

31. Since $\cos \alpha = \dfrac{b^2 + c^2 - a^2}{2bc}$ and substituting in the known values gives $\cos \alpha = \dfrac{(85.0)^2 + (105)^2 - (140)^2}{2(85.0)(105)}$ so $\alpha = 94.3°$. Solving for β and γ gives

$$\cos \beta = \frac{a^2 + b^2 - c^2}{2ac}$$
$$= \frac{(140)^2 + (105)^2 - (85.0)^2}{2(140)(105)}$$

and $\beta = 57.3°$. Then $\gamma = 180° - \alpha - \beta = 180° - 94.3° - 57.3° = 48.4°$.

33. Since the triangle is a right triangle, we can use the Pythagorean Theorem instead of the Law of Cosines, to find c. So $c^2 = (36.9)^2 + (20.45)^2$ and $c = 42.18$. To solve for α and β, we will use the Law of Cosines.

$$\cos \alpha = \frac{b^2 + c^2 - a^2}{2bc}$$
$$= \frac{(20.45)^2 + (42.18)^2 - (36.9)^2}{2(20.45)(42.18)}$$

so $\alpha = 61°$. Then $\gamma = 180° - 90° - 61° = 29°$.

35. Let x represent the distance between Buffalo and New York City. The angle opposite the side x is $49° + 9° = 58°$ and $x^2 = (210)^2 + (285)^2 - 2(210)(285) \cos 58°$ $x = 249$ mi.

37. Use $a^2 = b^2 + c^2 - 2bc \cos \alpha$ where $b = 5.20$, $c = 4.30$, and $\alpha = 68.4°$. Then $a^2 = (5.2)^2 + (3.4)^2 - 2(5.2)(3.4) \cos 68.4°$ and $a = 5.39$. The distance from the gun to the turret is 5.39 mi.

39. Choose points S and R across from P to form two right triangles, PSR and PQR. We know that $\alpha = 180° - 30.96° - 90° = 59.04°$ and that $\beta = 180° - \alpha = 180° - 59.04° = 120.96°$. Now we have SAS, so using the Law of Cosines gives $PQ^2 = 40^2 + 5^2 - 2(40)(5) \cos 120.96° = 1830.776$ so $PQ = 42.79$. The length of the wire running from P to Q is 42.79 ft.

41. Using $\cos\alpha = \dfrac{b^2 + c^2 - a^2}{2bc}$ and letting $a = 12$, $b = 8$, and $c = 10$ gives

$$\cos\alpha = \frac{(8)^2 + (10)^2 - (12)^2}{2(8)(10)}$$

and $\alpha = 82.8°$. Then solving for β using $\cos\beta = \dfrac{a^2 + b^2 - c^2}{2ac}$ gives

$$\cos\beta = \frac{(12)^2 + (10)^2 - (8)^2}{2(12)(10)} = 41.4°$$

Then $\gamma = 180° - 82.8° - 41.4° = 55.8°$. So, the left angle is $55.8°$ and the right angle is $41.4°$.

43. First solve for the three sides of the triangles. $DN = c = 1.875$, $NQ = a = 2.25$, and $QD = b = 2.125$. Then we can solve for the angles with $\cos\alpha = \dfrac{b^2 + c^2 - a^2}{2bc}$ we get

$$\cos\alpha = \frac{(2.125)^2 + (1.875)^2 - (2.25)^2}{2(2.125)(1.875)}$$

so $\alpha = 68.1°$, and using $\cos\beta = \dfrac{a^2 + b^2 - c^2}{2ac}$ we get

$$\cos\beta = \frac{(2.25)^2 + (1.875)^2 - (1.125)^2}{2(2.25)(1.875)}$$

so $\beta = 61.2°$, and then $\gamma = 180° - 68.1° - 61.2° = 50.7°$.

45. Since the circle has radius 5 in., we know that a radius connecting to a corner of the inscribed square has a length of 5 in. Subsequently, the diagonal of the square is 10 in. long. This is also the hypotenuse of a right, isoceles triangle. We can solve for one of the triangle's sides using the Pythagorean Theorem and $a^2 + a^2 = 10^2$ so $a = \sqrt{50}$. Since this is the angle of one side of the square (as well as one side of the triangle), we can multiply the result by 4 to find that the perimeter is $4\sqrt{50} = 20\sqrt{2}$ in. ≈ 28 in.

47. Since the radius of the circle is 783.4 ft, we know that a segment drawn to a vertex of the pentagon has length of 783.4. Two such adjacent lines in the pentagon form two legs of a triangle, with the remaining side as part of the perimeter of the pentagon. We can use the Law of Cosines to find the other side's length, given that the angle between the two adjacent segments is $\dfrac{360°}{5} = 72°$.

$$c^2 = a^2 + b^2 - 2ab \cos \gamma$$
$$= (783.4)^2 + (783.4)^2 - 2(783.4)(783.4) \cos 72° = 848134.044495$$

So $c = \sqrt{848134.044495} = 920.9$. When this answer is multiplied by 5, the perimeter of the Pentagon is 4604.7 ft.

49. Since $a^2 = 26^2 + 15^2 - 2(26)(15) \cos 15°$ then $a = 12.148$ mi. So at 15 mi/h and solving for minutes, it will take the boat $\frac{12.148}{15}(60) = 49$ min to reach Avalon.

51. The original heading was 210° (S30°W), and the correct original heading was 225°. The corrected heading after one hour is 244° (243.6374°).

53. Using $a^2 = 250^2 + 90^2 - 2(250)(90) \cos 10°$ gives $a \approx 162$ mi to travel and in hours the equation is $\dfrac{162}{\frac{250}{180} - .5} = 182.25$ mph. So the pilot must increase the airspeed to 182 mi/h.

55. Let B be in the standard position. Then $B(0,0)$, $C(a,0)$, and $A(c \cos \beta, c \sin \beta)$. If we find the distance from C to A,

$$b = \sqrt{(c \cos \beta - a)^2 + (c \sin \beta - 0)^2}$$
$$b^2 = (c \cos \beta - a)^2 + (c \sin \beta)^2$$
$$b^2 = c^2 \cos^2 \beta - 2ac \cos \beta + a^2 + c^2 \sin^2 \beta$$
$$b^2 = c^2(\cos^2 \beta + \sin^2 \beta) - 2ac \cos \beta + a^2$$
$$b^2 = c^2 + a^2 - 2ac \cos \beta$$

57. By the Law of Cosines, we know $\cos \gamma = \dfrac{a^2 + b^2 - c^2}{2ab}$ and $\cos \beta = \dfrac{a^2 + c^2 - b^2}{2ac}$, so $b \cos \gamma = \dfrac{a^2 + b^2 - c^2}{2a}$ and $c \cos \beta = \dfrac{a^2 + c^2 - b^2}{2a}$. Then, it holds that

$$b \cos \gamma + c \cos \beta = \dfrac{a^2 + b^2 - c^2}{2a} + \dfrac{a^2 + c^2 - b^2}{2a}$$
$$= \dfrac{2a^2}{2a} = a$$

59. By the Law of Cosines, we know $\cos\alpha = \dfrac{b^2 + c^2 - a^2}{2bc}$ and $\cos\beta = \dfrac{a^2 + c^2 - b^2}{2ac}$, so $b\cos\alpha = \dfrac{b^2+c^2-a^2}{2c}$ and $a\cos\beta = \dfrac{a^2+c^2-b^2}{2c}$. Then, it holds that

$$a\cos\beta + b\cos\alpha = \frac{a^2 + c^2 - b^2}{2c} + \frac{b^2 + c^2 - a^2}{2c}$$
$$= \frac{2c^2}{2c} = c$$

61. Given that $\cos\beta = \dfrac{a^2 + c^2 - b^2}{2ac}$ we can see that $\cos\beta + 1 = \dfrac{a^2 + c^2 - b^2 + 2ac}{2ac}$. Also, we can find that

$$(a + c - b)(a + b + c) = a^2 + ab + ac + ca + cb + c^2 - ba - b^2 - cb$$
$$= (a^2 + c^2 - b^2 + 2ac)$$

Thus, $\cos\beta + 1 = \dfrac{(a + c - b)(a + b + c)}{2ac}$.

Problem Set 5.2

5. one solution; 2 significant digits

7. two solutions; 2 significant digits

9. one solution; 3 significant digits

11. one solution; 2 significant digits

13. two solution; 2 significant digits

15. no solution; 2 significant digits

17. This triangle cannot be solved since the sum of the angles is not less than 180°.

19. Since $\gamma = 180° - 25° - 110° = 45°$. Then if $\dfrac{\sin 25°}{a} = \dfrac{\sin 110°}{23}$ then $a = 10$, and if $\dfrac{\sin 110°}{23} = \dfrac{\sin 45°}{c}$ then $c = 17$.

21. Since $\alpha = 180° - 85° - 25° = 70°$. Then if $\dfrac{\sin 70°}{a} = \dfrac{\sin 85°}{90}$ then $a = 85$, and if $\dfrac{\sin 25°}{c} = \dfrac{\sin 85°}{90}$ then $c = 38$.

23. Since $\beta = 180° - 120° - 7° = 53°$. Then if $\dfrac{\sin 7°}{43} = \dfrac{\sin 120°}{a}$ then $a = 310$, and if $\dfrac{\sin 7°}{43} = \dfrac{\sin 53°}{b}$ then $b = 280$.

25. Since $\gamma = 180° - 18.3° - 54.0° = 107.7°$. Then if $\dfrac{\sin 18.3°}{107} = \dfrac{\sin 54°}{b}$ then $b = 276$, and if $\dfrac{\sin 18.3°}{107} = \dfrac{\sin 107.7°}{c}$ then $c = 325$.

27. Since $\dfrac{\sin \alpha}{10.8} = \dfrac{\sin 21.9°}{8.80}$ then $\alpha = 27.2°$. Then using $\gamma = 180° - \alpha - \beta$ gives $\gamma = 180° - 27.2° - 21.9° = 130.9°$. Then if $\dfrac{\sin 130.9°}{c} = \dfrac{\sin 21.9°}{8.80}$, $c = 17.8$. However, $\alpha = 152.8°$ also, and thus $\gamma = 180° - 21.9° - 152.8° = 5.3°$ and then we know that $\dfrac{c}{\sin 5.3°} = \dfrac{8.8}{\sin 21.9°}$ and consequently, $c = \dfrac{8.8 \sin 5.3°}{\sin 21.9°} = 2.2$.

29. Solving for a, $a^2 = (55.0)^2 + (92.0)^2 - 2(55.0)(92.0) \cos 98°$ and $a = 114$. Then $\dfrac{\sin 98°}{114} = \dfrac{\sin \beta}{55}$ so $\beta = 28.7°$. With $\gamma = 180° - \alpha - \beta$ we have $\gamma = 180° - 98.0° - 28.7° = 53.3°$.

31. Using $\gamma = 180° - \alpha - \beta$ we have $\gamma = 180° - 85.2° - 38.7° = 56.1°$. Then solving for a gives $\dfrac{\sin 56.1°}{123} = \dfrac{\sin 85.2°}{a}$ and $a = 147.7$. If $\dfrac{\sin 56.1°}{123} = \dfrac{\sin 38.7°}{b}$ then $b = 92.7$.

33. Cannot solve the triangle with the given information. The Laws of Cosines and Sines both require at least one side of the triangle before they can be applied.

35. Cannot solve the triangle because the sum of the length of the sides of the triangle are less than the hypotenuse, so the sides given do not form a triangle.

37. Solving for γ, $\dfrac{\sin 25°}{1.1} = \dfrac{\sin \gamma}{2.1}$ and $\gamma = 54°$. Using $\beta = 180° - \alpha - \gamma$ gives $\gamma = 180° - 25° - 54° = 101°$. Then if $\dfrac{\sin 25°}{1.1} = \dfrac{\sin 101°}{b}$ and $b = 2.6$. Also there is a second triangle with $\beta = 29°$, $\gamma = 126°$, and $b = 1.3$.

39. Let one acute angle be $72°$ and a second acute angle at the top of the triangle is $90° - 72° = 18°$. Let y be the height of the building, then $\dfrac{y}{\sin 52°} = \dfrac{100}{\sin 18°}$ and $y = 255$. So, the height of the building is approximately 260 ft.

41. Using the Law of Sines, we can see that $\dfrac{150}{\sin 14°} = \dfrac{x}{\sin 28°}$ and thus $x = \dfrac{150 \sin 28°}{\sin 14°} = 291$. The distance to the target is 291 km.

43. The aircraft is 4318 m from the first observation point and 6815 m from the second.

45. From Problem 44, with l as the length of the tower, the unknown angle is $\theta = 180 - 84.65 - 20.24 = 75.11$. Then $\dfrac{\sin 20.24}{l} = \dfrac{\sin 75.11}{500}$, and $l = 178.986$. If we let $D^2 = (500)^2 + l^2 - 2l(500) \cos 95.45°$, then $D = 546.841$. So $\dfrac{\sin \gamma}{178.986} = \dfrac{\sin 95.45}{546.841}$ gives $\gamma = 19.0°$ which is the angle of elevation.

47. If $\alpha = 180° - 34.06°$ then $\alpha = 145.94°$ and if $\beta = 180° - \alpha - 27.77°$ then $\beta = 6.29°$. Using the Law of Sines to solve for the side ST gives $\dfrac{\sin 27.77°}{ST} = \dfrac{\sin 6.29°}{1000}$ so $ST = 4253$ and the length of the ski lift is 4253 ft long.

49. Since
$$\cos \alpha = \frac{(1048)^2 + (3950)^2 - (3950)^2}{2(1048)(3950)}$$

Then $\alpha = 82.37°$. Also, α and β are supplementary, so $180 - \alpha = \beta$, and $\beta = 97.6°$. This means that $\gamma = 180° - 70.9° - 97.6° = 11.5°$. Then we can use Law of Sines to determine the distance x and $\dfrac{x}{\sin 11.5°} = \dfrac{104.8}{\sin 70.9°}$ and consequently, $x = 220$. *Sputnik* is 220 mi above point B.

51. Drop a perpendicular from C and let h be the length of the perpendicular. Then $\sin \alpha = \frac{h}{b}$ and $\sin \beta = \frac{h}{a}$. So, $h = b \sin \alpha$ and $h = a \sin \beta$. Thus, $b \sin \alpha = a \sin \beta$. Dividing by ab, gives $\frac{b \sin \alpha}{ab} = \frac{a \sin \beta}{ab}$ and $\frac{\sin \alpha}{a} = \frac{\sin \beta}{b}$.

53.

$$\frac{\sin \alpha}{a} = \frac{\sin \beta}{b}$$

$$b \sin \alpha = a \sin \beta$$

$$\frac{b \sin \alpha}{b \sin \beta} = \frac{a \sin \beta}{b \sin \beta}$$

$$\frac{\sin \alpha}{\sin \beta} = \frac{a}{b}$$

55.

$$\frac{2 \cos \frac{1}{2}(\alpha + \beta) \sin \frac{1}{2}(\alpha - \beta)}{2 \sin \frac{1}{2}(\alpha + \beta) \cos \frac{1}{2}(\alpha - \beta)} = \frac{\sin \alpha - \sin \beta}{\sin \alpha + \sin \beta} \quad \text{Addition laws}$$

$$= \frac{a - b}{a + b}$$

57. From problem 56,

$$\frac{\tan \frac{1}{2}(\alpha - \beta)}{\tan \frac{1}{2}(\alpha + \beta)} = \frac{a - b}{a + b}$$

Let $\alpha = \beta$, $\beta = \gamma$, $a = b$, and $b = c$, then

$$\frac{\tan \frac{1}{2}(\beta - \gamma)}{\tan \frac{1}{2}(\beta + \gamma)} = \frac{b - c}{b + c}$$

59. We know that

$$\frac{a}{c} = \frac{\sin\alpha}{\sin\gamma} \quad \text{and} \quad \frac{b}{c} = \frac{\sin\beta}{\sin\gamma}$$

From the Law of Sines,

$$\frac{a}{c} - \frac{b}{c} = \frac{\sin\alpha}{\sin\gamma} - \frac{\sin\beta}{\sin\gamma}$$

$$\frac{a-b}{c} = \frac{\sin\alpha - \sin\beta}{\sin\gamma}$$

$$= \frac{2\cos\frac{1}{2}(\alpha+\beta)\sin\frac{1}{2}(\alpha-\beta)}{2\sin(\frac{1}{2}\gamma)\cos(\frac{1}{2}\gamma)}$$

Since $\alpha + \beta + \gamma = 180°$, we know that $\frac{1}{2}(\alpha+\beta) + \frac{1}{2}\gamma = 90°$. Thus, $\frac{1}{2}(\alpha+\beta)$ and $\frac{1}{2}\gamma$ are complementary, so that $\sin\left(\frac{1}{2}\gamma\right) = \cos\left(\frac{1}{2}(\alpha+\beta)\right)$. Thus,

$$\frac{a-b}{c} = \frac{\sin\left(\frac{1}{2}(\alpha-\beta)\right)}{\cos\left(\frac{1}{2}\gamma\right)}$$

61. With $\sin\gamma = \dfrac{w}{h}$ and $\sin\phi = \dfrac{W}{h}$ and eliminating h, we have $\dfrac{\sin\gamma}{w} = \dfrac{\sin\phi}{W}$ and $W\sin\gamma = w\sin\phi$. Then, since $\gamma + \phi = \theta$, substitute $\theta - \phi = \gamma$ and solve for ϕ.

$$W[\sin(\theta - \phi)] = w\sin\phi$$

$$W[\sin\theta\cos\phi - \cos\theta\sin\phi] = w\sin\phi$$

$$W\sin\theta\cos\phi = w\sin\phi + W\cos\theta\sin\phi$$

$$= (w + W\cos\theta)\sin\phi$$

$$\frac{W\sin\theta\cos\phi}{(w + W\cos\theta)\cos\phi} = \frac{(w + W\cos\theta)\sin\phi}{(w + W\cos\theta)\cos\phi}$$

$$\frac{W\sin\theta}{w + W\cos\theta} = \tan\phi$$

$$\phi = \tan^{-1}\left(\frac{\sin\theta}{\frac{w}{W} + \cos\theta}\right)$$

To obtain γ, we calculate either $\theta - \phi$, or with symmetry we have

$$\gamma = \tan^{-1}\left(\frac{\sin\theta}{\frac{W}{w} + \cos\theta}\right)$$

Problem Set 5.3

1. $K = \frac{1}{2}bh$

3. $K = \frac{1}{2}bc \sin \alpha$

5. Use the Law of Cosines to find an included angle and then use the formulas for two angles and an included side.

7. $K = \frac{1}{2}\theta r^2$

9. $K = \frac{1}{2}bh = \frac{1}{2}(15.6)(2.51) = 19.58$ sq units

11. $K = 13.6$ sq units

13. $K = 44$ sq units

15. $K = \frac{1}{2}ab \sin \gamma = \frac{1}{2}(15)(8) \sin 38° = 37$ sq units

17. $K = \frac{1}{2}bc \sin \alpha = \frac{1}{2}(14)(12) \sin 82° = 83$ sq units

19. Using $\alpha = 180° - \beta - \gamma$ gives $\alpha = 180° - 50° - 100° = 30°$. Then $K = \dfrac{30^2 \sin 50° \sin 100°}{2 \sin 30°} = $ 680 sq units.

21. Using $\beta = 180° - 50° - 60°$ gives $\beta = 70°$. Then $K = \dfrac{40^2 \sin 50° \sin 60°}{2 \sin 70°} = 570$ sq units.

23. $s = \frac{1}{2}(a + b + c) = \frac{1}{2}(15) = \frac{15}{2}$, $K = \sqrt{\frac{15}{2}\left(\frac{15}{2} - 5\right)\left(\frac{15}{2} - 5\right)\left(\frac{15}{2} - 5\right)}$, then $K \approx 10.8$ sq units.

25. $s = \frac{1}{2}(7 + 8 + 2) = \frac{17}{2}$, $K = \sqrt{\frac{17}{2}(\frac{17}{2} - 7)(\frac{17}{2} - 8)(\frac{17}{2} - 2)}$, and $K = 6.4$ sq units.

27. $s = \frac{1}{2}(11 + 9 + 8) = 14$, $K = \sqrt{14(14 - 11)(14 - 9)(14 - 8)}$, and $K = 35$ sq units.

29. Using $a^2 = b^2 + c^2 - 2bc \cos \alpha$ gives $12^2 = 9^2 + c^2 - 18c \cos 52°$. Then solving for c, $c^2 - 18 \cos 52° c - 63 = 0$ using the quadratic equation gives $c = 15.220941$. $K = \frac{1}{2}(9.00)(15.220941) \sin 52° = 54.0$ sq units.

31. Using $a^2 = b^2 + c^2 - 2bc \cos \alpha$ gives $10.2^2 = 11.8^2 + c^2 - 2(11.8)c \cos 47°$. Then solving for c, $c^2 - 23.6 \cos 47° c + 35.2 = 0$ gives $c = 13.484818$ and $c = 2.6103429$. So, there are two triangles to solve. First using $c = 13.484818$, $K = \frac{1}{2}(11.8)(13.484818) \sin 47° = 58.2$ sq units. Then using $c = 2.6103429$, $K = \frac{1}{2}(11.8)(2.6103429) \sin 47° = 11.3$ sq units.

33. Using $c^2 = a^2 + b^2 - 2ab \cos \gamma$ gives $52.2^2 = a^2 + 82.5^2 - 2(82.5)a \cos 32.1°$. Then solving for a, $a^2 - 165 \cos 32.1° a + 4081.41 = 0$ and $a = 98.222358$ and $a = 41.552759$. So, there are two triangles to solve. First using $a = 98.222358$, $K = \frac{1}{2}(98.222358)(82.5) \sin 32.1° = 2153$ sq units. Then using $c = 41.552759$, $K = \frac{1}{2}(41.552759)(82.5) \sin 32.1° = 911$ sq units.

35. Using $\alpha = 180° - \beta - \gamma$ gives $\alpha = 180° - 18° - 15° = 147°$. Then $K = \dfrac{b^2 \sin \alpha \sin \gamma}{2 \sin \beta}$ and

$$K = \frac{23.5^2 \sin 147° \sin 18°}{2 \sin 15°} = 180 \text{ sq units}.$$

37. No solution because the angles given add to $180°$, so no third angle exists and thus, no triangle is formed.

39. $s = \frac{1}{2}(124 + 325 + 351) = 400$ and $K = \sqrt{(400)(400 - 124)(400 - 325)(400 - 351)} = 20142$, rounded to three significant digits is 20100 sq units.

41. $K = \frac{1}{2}\theta r^2 = \frac{1}{2}(1.0)(12)^2 = 72 \text{ in.}^2$

43. $K = \frac{1}{2}\theta r^2 = \frac{1}{2}(2.3)(3.5)^2 = 14.1 \text{ m}^2$

45. Since $30° = \frac{\pi}{6}$, $K = \frac{1}{2}\theta r^2 = \frac{1}{2}(\frac{\pi}{6})(10)^2 = 26 \text{ in.}^2$

47. Connect the center of the circle to the three vertices of the triangle. One of the central angles formed by two such segments is measured at $\frac{1}{3}(360°) = 120°$. Then $K = \frac{1}{2}(1)(1)\sin 120° = \frac{\sqrt{3}}{4}$. Since there are three triangles with such an area, we find the total area to be $-3\frac{\sqrt{3}}{4}$ or $\frac{3\sqrt{3}}{4}$. The answer to two decimal places is approximately 1.30 in.^2

49. Connect the center of the circle to the five vertices of the triangle. One of the central angles formed by two such adjacent segments is measured at $\frac{1}{5}(360°) = 72°$. Then $K = \frac{1}{2}(1)(1)\sin 72° = \dfrac{\sin 72°}{2}$. Since there are five triangles with such an area, we find the total area to be $5\left(\dfrac{\sin 72°}{2}\right)$ or $\frac{5}{2}\sin 72°$. The answer to two decimal places is approximately 2.38 in.^{2}

51. a. θ in radian measure is $\dfrac{\overline{AC}}{\overline{OA}} = \dfrac{\overline{AC}}{1} = \overline{AC}$ and the area of $\triangle AOD = \frac{1}{2}\overline{OD}\,\overline{AD}$ is $K = \frac{1}{2}\sin\theta\cos\theta = \frac{1}{4}\sin 2\theta$.

   b.   Since $\overline{BC} = \tan\theta$, so the area of $\triangle BOC$ is $K = \frac{1}{2}\tan\theta$.

   c. $K = \frac{1}{2}\theta$.

53. Let x be the apothem and using $\dfrac{\sin 90}{1} = \dfrac{\sin 54}{x} = 0.81$. So, the apothem is 0.81 in.

55. $K = \frac{1}{2}\theta r^2$ and $r = \sqrt{\dfrac{2K}{\theta}} = \sqrt{\dfrac{2(54.4)}{1.78}} = 7.82$ cm.

57. The total is 220447 ft^2 which is approximately 5.06 acres. The cost is about \$ 225.

59. Use $V = \frac{1}{3}\pi r^3 \tan\alpha$ where $\alpha = 36°$ and $r = 30$ ft. Then $V = \frac{1}{3}\pi(30)^3\tan 36° = 20543$. So, rounded to two places of accuracy, $V = 21,000 \text{ ft}^3$. Since cubic yards are required, convert using $1 \text{ yd}^3 = 27 \text{ ft}^3$ and $V = (20543)\frac{1}{27} = 760 \text{ yd}^3$.

61.   a.

$$V = h(\text{Area of base})$$
$$= h[(\text{Area of sector} - \text{Area of triangle})]$$
$$= h\left[\frac{1}{2}\theta r^2 - r^2 \sin\frac{\theta}{2}\cos\frac{\theta}{2}\right]$$
$$= hr^2\left(\frac{\theta}{2} - \sin\frac{\theta}{2}\cos\frac{\theta}{2}\right)$$
$$b = r\sin\frac{\theta}{2}$$
$$h = r\cos\frac{\theta}{2}$$

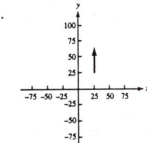

Area of the triangle is $2\left(\frac{1}{2}bh\right) = r^2 \sin\frac{\theta}{2}\cos\frac{\theta}{2}$.

b. $V = (1296)3^2\left(\sin\frac{60}{2}\cos\frac{60}{2}\right) = 117$ in.^{3}

Problem Set 5.4

5.

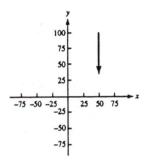

7.

9.

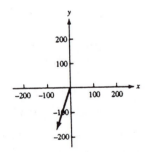

11.

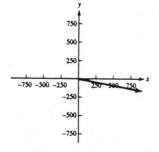

13.

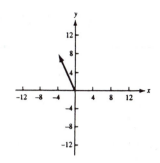

15.

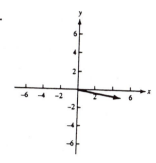

17. $|v| = \sqrt{3^2 + 4^2} = 5$ and $\tan\theta = \frac{3.0}{4.0}$, so $\theta = 37°$. Resultant has magnitude of 5.0 and direction S37°W.

19. $|v| = \sqrt{100^2 + 200^2} = 223.60679$ and $\tan\theta = \frac{200}{100}$, so $\theta = 63°$. Resultant has magnitude of 220 and direction 110°.

21. $|v| = \sqrt{75^2 + 200^2 - 2(75)(200)\cos 50°} = 162$ and $\sin\theta = \dfrac{|b|\sin(180° - \phi)}{|v|}$ so $\sin\theta = \dfrac{200\sin 50°}{162}$ and $\theta = 71°$. Resultant has magnitude of 162 and heading 220°.

23. $\cos 60° = \dfrac{V_x}{20}$ and $\sin 60° = \dfrac{V_y}{20}$, so $V_x = 10$ and $V_y = 10\sqrt{3} \approx 17$.

25. $\cos 27° = \dfrac{V_x}{125}$ and $\sin 27° = \dfrac{V_y}{125}$, so $V_x = 110$ and $V_y = 57$.

27. $\cos 130° = \dfrac{V_x}{525}$ and $\sin 130° = \dfrac{V_y}{525}$, so $V_x = 337$ and $V_y = 402$.

29. $v = \sqrt{43^2 + 240^2} = 244$ and $\tan\theta = \frac{43}{240}$ so $\theta = 10°$. Then the speed of the plane is 244 mi/h with a course of 260°.

31. $\cos 17.8° = \dfrac{V_x}{2120}$ and $\sin 17.8° = \dfrac{V_y}{2120}$ so $V_x = 2019$ and $V_y = 648$. The vertical component is 648 ft/s and the horizontal component is 2019 ft/s.

33. In two hours, the plane has traveled 496 mi. So, $\sin 43° = \frac{V_y}{496}$ and $V_y = 338$. Therefore, the plane has traveled 338 mi east in two hours.

35. Using the Law of Cosines, $\cos\alpha = \dfrac{70^2 + 40^2 - 50^2}{2(70)(40)}$ and $\alpha = 44°$; $\cos\beta = \dfrac{50^2 + 40^2 - 70^2}{2(50)(40)}$ and $\beta = 102°$; $\gamma = 180° - 44° - 102°$ and $\gamma = 34°$; $\theta = 180° - 34° = 146°$.

37. Using the Law of Cosines, $12^2 = 15^2 + c^2 - 2(15)c\cos 37°$ and $c^2 - (30c)\cos 37° + 81 = 0$ solving for c gives $c = \dfrac{30\cos 37° \pm \sqrt{(-30\cos 37°)^2 - 4(1)(81)}}{2}$ and $c = 19.885809$ and $c = 4.0732565$. So rounding to two decimal places of accuracy, the third side of $x' = 20$ km or $x = 4.1$ km. Since the boat's rate is 10 km/hr, using (rate)(time) = distance gives $t = \frac{20}{10} = 2$ h or $t = \frac{4.1}{10} = .41$ h.

39. Using $v = \sqrt{a^2 + b^2 - 2ab\cos(180° - \phi)}$ we have $v = \sqrt{200^2 + 100^2 - 2(200)(100)\cos(180° - 35°)} = 288$. So solving for both 100-lb force and 200-lb force, $\sin\theta = \dfrac{100\sin(180° - 35°)}{288}$ so $\theta = 11.5°$ and $\sin\theta = \dfrac{200\sin(180° - 35°)}{288}$ so $\theta = 23.5°$ The force has magnitude 288 lb and the tree moves in a path at an angle of $23.5°$ from the 200-lb source and an angle of $11.5°$ with the 100-lb source.

41. The angle is $68° - 16° = 52°$ and $|v| = \sqrt{12^2 + 25^2 - 2(12)(25)\cos(180° - 52°)} = 33.7$ and $\sin\theta = \dfrac{25\sin(180° - 52°)}{33.7}$ so $\theta = 35.7°$. The boat is travelling 33.7 knots at $68° - 35.7°$, that is, at N32.3°W.

43. The angle inside the triangle at the common tail of the two given forces is $360° - 280° + 35° = 115°$ or $80° + 35° = 115°$. Let x be the distance between the heads of the two given vectors. Using SAS, $x^2 = 180^2 + 260^2 - 2(180)(260)\cos 115°$ and $x = 373.6$. After one hour the planes are about 374 mi apart.

45. Let x respresent the distance between the heads of the two given vectors. Since 2 hours have elapsed, the vectors have magnitudes of 48.2 miles and 19.6 miles. The angle opposite x is $42.1° + 90° + (90° - 58.5°) = 163.6°$. Using SAS, we have $x^2 = 19.6^2 + 48.2^2 - 2(19.6)(48.2)\cos 163.6°$ and $x = 67.2$. Thus, the boats are 67.2 miles apart after 2 hours.

47. The tension on $\overline{AC}$ is 326 lb, and on $\overline{BC}$ the tension is 321 lb.

49. Since $\dfrac{\sin 12°}{x} = \dfrac{\sin 90°}{250}$, $x = 52$-lb force is needed to keep the barrel from sliding down the incline. The barrel is pushing against the plane with a force of $\dfrac{\sin 90°}{250} = \dfrac{\sin 78°}{y}$ or $y = 240$ lb.

51. $\dfrac{\sin 90°}{800} = \dfrac{\sin\alpha}{486}$ so $\alpha = 37.4°$. The angle of inclination is $37.4°$.

53. $\dfrac{\sin 90°}{x} = \dfrac{\sin 18.2°}{5250}$ so $x = 16808$. The heaviest cargo that can be pulled can weigh 16808 lb.

55. We know by Law of Sines that $\dfrac{x}{\sin 40°} = \dfrac{53.5}{\sin 50°}$ and then $x = 44.9$. Similarly, we can compute $\dfrac{y}{\sin 90°} = \dfrac{53.5}{\sin 50°}$ and $y = 69.8$. The tension of the cable is 69.8 lb. The horizontal force is 44.9 lb.

57. We know that $W\cos\theta = \frac{1}{6}W$. Consequently, $\cos\theta = \frac{1}{6}$. Thus, $\theta \approx 80.4°$.

59. The track should be banked at an angle of $77°$.

Problem Set 5.5

9. $a = 8\cos 30° = 4\sqrt{3}$; $b = 8\sin 30° = 4$; $\mathbf{v} = 4\mathbf{i}\sqrt{3} + 4\mathbf{j}$

11. $a = 5\sqrt{2}\cos 315° = 5$; $b = 5\sqrt{2}\sin 315° = -5$; $\mathbf{v} = 5\mathbf{i} - 5\mathbf{j}$

13. $a = 10.4813 \cos 214.5° = -8.6379$; $b = 10.4813 \sin 214.5° = -5.9367$; $\mathbf{v} = -8.6379\mathbf{i} - 5.9367\mathbf{j}$

15. $\langle AB \rangle = \langle 4 - (-1), 5 - (-2) \rangle = \langle 5, 7 \rangle$; $\mathbf{v} = 5\mathbf{i} + 7\mathbf{j}$

17. $\langle AB \rangle = \langle (-8) - 5, -3 - (-3) \rangle = \langle -13, 0 \rangle$; $\mathbf{v} = 13\mathbf{i} + 0\mathbf{j}$

19. $\langle AB \rangle = \langle 4 - 0, 1250 - 0 \rangle = \langle 4, 1250 \rangle$; $\mathbf{v} = 4\mathbf{i} + 1250\mathbf{j}$

21. $|\mathbf{v}| = \sqrt{5^2 + (-12)^2} = 13$

23. $|\mathbf{v}| = \sqrt{(-3)^2 + 5^2} = \sqrt{34}$

25. $|\mathbf{v}| = \sqrt{5^2 + (-8)^2} = \sqrt{89}$

27. $|\mathbf{v}| = \sqrt{(6\sqrt{2}^2 + (-5\sqrt{2})^2} = \sqrt{122}$

29. $\mathbf{v} \cdot \mathbf{w} = (2)(6) + (3)(-9) = -15$; not orthogonal

31. $\mathbf{v} \cdot \mathbf{w} = (5)(8) + (4)(-10) = 0$; orthogonal

33. $\mathbf{v} \cdot \mathbf{w} = (\cos 20°)(\cos 70°) + (\sin 20°)(\sin 70°) = .64278$; not orthogonal

35. $\mathbf{v} \cdot \mathbf{w} = (8)(-5) + (-6)(12) = -112$; $|\mathbf{v}| = \sqrt{8^2 + (-6)^2} = 10$; $|\mathbf{w}| = \sqrt{(-5)^2 + 12^2} = 13$; $\cos\theta = \dfrac{\mathbf{v} \cdot \mathbf{w}}{|\mathbf{v}||\mathbf{w}|} = \dfrac{-112}{130}$ where $\theta = 149°$.

37. $\mathbf{v} \cdot \mathbf{w} = (7)(2\sqrt{15}) - (\sqrt{15})(14) = 0$; $|\mathbf{v}| = \sqrt{7^2 + (\sqrt{15})^2} = 8$; $|\mathbf{w}| = \sqrt{(2\sqrt{15})^2 + 14^2} = 16$; $\cos\theta = \dfrac{\mathbf{v} \cdot \mathbf{w}}{|\mathbf{v}||\mathbf{w}|} = 0$ where $\theta = 90°$.

39. $\mathbf{v} \cdot \mathbf{w} = (3)(2) + (9)(-5) = -39$; $|\mathbf{v}| = \sqrt{3^2 + 9^2} = 3\sqrt{10}$; $|\mathbf{w}| = \sqrt{2^2 + (-5)^2} = \sqrt{29}$; $\cos\theta = \dfrac{\mathbf{v} \cdot \mathbf{w}}{|\mathbf{v}||\mathbf{w}|} = \dfrac{-39}{\sqrt{90}\sqrt{29}}$ where $\theta = 140°$.

41. $\mathbf{v} \cdot \mathbf{w} = (0)(0) + (1)(1) = 1$; $|\mathbf{v}| = \sqrt{0^2 + 1^2} = 1$; $|\mathbf{w}| = \sqrt{0^2 + 1^2} = 1$; $\cos\theta = \dfrac{\mathbf{v} \cdot \mathbf{w}}{|\mathbf{v}||\mathbf{w}|} = 1$ where $\theta = 0°$.

43. $\mathbf{v} \cdot \mathbf{w} = (0)(-1) + (-1)(0) = 0$; $|\mathbf{v}| = \sqrt{0^2 + (-1)^2} = 1$; $|\mathbf{w}| = \sqrt{(-1)^2 + 0^2} = 1$; $\cos\theta = \dfrac{\mathbf{v} \cdot \mathbf{w}}{|\mathbf{v}||\mathbf{w}|} = 0$ where $\theta = 90°$.

45. $\mathbf{v} \cdot \mathbf{w} = (\sqrt{2})(\frac{\sqrt{3}}{2}) + (-\sqrt{2})(\frac{\sqrt{3}}{2}) = 0$; $\cos\theta = \dfrac{\mathbf{v} \cdot \mathbf{w}}{|\mathbf{v}||\mathbf{w}|} = 0$ where $\theta = 75°$. We have no need to check the magnitudes of $\mathbf{v}$ and $\mathbf{w}$ since $\mathbf{v} \cdot \mathbf{w} = 0$.

47. $\mathbf{v} \cdot \mathbf{w} = (-1)(-2\sqrt{2}) + (0)(2\sqrt{2}) = 2\sqrt{2}$; $|\mathbf{v}| = \sqrt{(-1)^2 + 0^2} = 1$; $|\mathbf{w}| = \sqrt{(2\sqrt{2})^2 + (2\sqrt{2})^2} = 4$; $\cos\theta = \dfrac{\mathbf{v} \cdot \mathbf{w}}{|\mathbf{v}||\mathbf{w}|} = \dfrac{\sqrt{2}}{2}$ where $\theta = 45°$.

49. $\mathbf{v} \cdot \mathbf{w} = (-3)(6) + (2)(9) = 0$; $\cos\theta = \dfrac{\mathbf{v} \cdot \mathbf{w}}{|\mathbf{v}||\mathbf{w}|} = 0$ where $\theta = 90°$.

51. $\mathbf{v} \cdot \mathbf{w} = (-2)(4) + (-a)(5) = 0$ and solving for a gives $a = \dfrac{-8}{5}$.

53. $\mathbf{v} \cdot \mathbf{w} = (3)(-6) + (a)(a) = 0$ and solving for a gives $a = \pm\sqrt{18}$.

55. Using $\mathbf{W} = \mathbf{F} \cdot \mathbf{d}$ and $\mathbf{d} = 4\mathbf{i} + 3\mathbf{j}$, we have $\mathbf{W} = (4)(4) + (5)(3) = 31$ ft-lb.

57. Using $\mathbf{W} = \mathbf{F} \cdot \mathbf{d}$ and $\mathbf{d} = -6\mathbf{i} + 8\mathbf{j}$, we have $\mathbf{W} = (\sqrt{2})(-6) + (4)(8) = -6\sqrt{2} + 32$ ft-lb.

59. Let $\mathbf{F}_1 = |\mathbf{F}_1|(\cos 31°\mathbf{i} + \sin 31°\mathbf{j})$ and $\mathbf{F}_2 = |\mathbf{F}_2|(\cos 37°\mathbf{i} + \sin 37°\mathbf{j})$ then solving for the work $\mathbf{W} = (|\mathbf{F}_1| + |\mathbf{F}_2|)[(\cos 31° + \cos 37°)\mathbf{i} + (\sin 31° + \sin 37°)\mathbf{j}]$

61. $\mathbf{W} = (165)(5) = 325$ ft-lb

63. $\mathbf{W} = |\mathbf{F}|d = (80)(30) = 2400$ ft-lb

65. First we compute the force pulling the child horizontally. $\mathbf{F} = 25\cos 46° = 17$ lb of force pulling the child directly forward. Then solving for the work exerted $\mathbf{W} = |\mathbf{F}|d = (17)(150) = 2600$ ft-lb to pull the wagon.

67. The force in the north direction is $500\cos 30° = 433.01270$ lbs. Therefore the work is $(433.01270)(100) \approx 43301$ ft-lb.

69. Let $\mathbf{v} = a\mathbf{i} + b\mathbf{j}$, $\mathbf{w} = c\mathbf{i} + d\mathbf{j}$, and $(\mathbf{v} + \mathbf{w}) = (a + c)\mathbf{i} + (b + d)\mathbf{j}$. Then

$$
\begin{aligned}
(\mathbf{v} + \mathbf{w}) \cdot (\mathbf{v} + \mathbf{w}) &= (a + c)^2 + (b + d)^2 \\
&= a^2 + 2ac + c^2 + b^2 + 2bd + d^2 \\
&= (a^2 + b^2) + (c^2 + d^2) + 2(ac + bd) \\
&= |\mathbf{v}|^2 + |\mathbf{w}|^2 + 2(\mathbf{v} \cdot \mathbf{w})
\end{aligned}
$$

Chapter 5 Sample Test

1.   a. $a^2 = b^2 + c^2 - 2bc\cos\alpha$ b. $\cos\beta = \dfrac{a^2 + c^2 - b^2}{2ac}$

    c. $\dfrac{\sin\alpha}{a} = \dfrac{\sin\beta}{b} = \dfrac{\sin\gamma}{c}$

3. $a = 19$; $b = 7.2$; $c = 15$; $\alpha = 113°$; $\beta = 20°$; $\gamma = 47°$

5.   Area is $\frac{1}{2}(16)(43)\sin 113° \approx 320$.

7. $s = \frac{1}{2}(121 + 46 + 92) = 129.5$ and $K = \sqrt{129.5(8.5)(83.5)(37.5)} \approx 1900$.

9. a. $x = \sqrt{2}\cos 45°, y = \sqrt{2}\sin 45°, \mathbf{v} = \mathbf{i} + \mathbf{j}$

   b. $x = 9.3\cos 118°, y = 9.3\sin 118°, \mathbf{v} = -4.366\mathbf{i} + 8.211\mathbf{j}$

   c. The horizontal x distance between the two points is $1 - (-5) = 6$, and the vertical y distance between the two points is $-2 - (-7) = 5$. Thus, $\mathbf{v} = 6\mathbf{i} + 5\mathbf{j}$.

11. Since $\gamma = 180° - 105.8° - 46.5° = 27.7°$ and using the Law of Sines to compute the remaining sides of the garden gives $\dfrac{\sin 46.5°}{38} = \dfrac{\sin 105.8°}{x}$, so $x = 50.4$, and $\dfrac{\sin 46.5°}{38} = \dfrac{\sin 27.7°}{x}$ so $x = 24.4$. Add the sides to get $38 + 50.4 + 24.4 = 112.8$ ft of fencing.

13. We need to find the distance, x, along the westward path. To accomplish this, we need to solve the triangle using the Law of Cosines, $6^2 = 7^2 + c^2 - 2(7)\cos 40°$ and solving for $c = \dfrac{14\cos 40° \pm \sqrt{(14\cos 40)^2 - 4(1)(13)}}{2}$.

    Then $c = 9.331489$ and $c = 1.393132$. At 2 mi/h the trip will take $\dfrac{9.331489}{2}(60) \approx 279$ min and $\dfrac{1.393132}{2}(60) \approx 42$ min. So you will be six miles from Ferndale at 4:40 P.M. or at 12:42 P.M.

15. Use the Law of Cosines. Let x be the distance between the entrance and the exit then $x^2 = 382^2 + 485^2 - 2(382)(485)(\cos 58°)$ so $x \approx 429$ ft.

Chapter 5 Miscellaneous Problems

1. a. $A(\cos\theta, 0); P(\cos\theta, \sin\theta)$ b. $|OA| = \cos\theta$

   c. $|PA| = \sin\theta$ d. $|OB| = \cos^2\theta$ e. $K = \frac{1}{2}\cos^3\theta\sin\theta$

3. a. $A(0, \sin\theta); P(\cos\theta, \sin\theta)$ b. $|OA| = \sin\theta$

   c. $|PA| = \cos\theta$ d. $|OB| = \sin^2\theta$ e. $K = \frac{1}{2}\sin^3\theta\cos\theta$

5. a. $\mathbb{R}$ b. $-1 \le y \le 1$ c. amplitude is 1; period is 2π

   d. 1   e. It crosses the x-axis when $x = n\pi$.

7. not an identity; let $\theta = \frac{\pi}{4}$, then $\sin^2\frac{\pi}{4} - \sin\frac{\pi}{4} - 1 \neq 0$.

9. identity;

$$
\begin{aligned}
\tan\theta + \cot\theta &= \frac{\sin\theta}{\cos\theta} + \frac{\cos\theta}{\sin\theta} \\
&= \frac{\sin^2\theta + \cos^2\theta}{\cos\theta\sin\theta} \\
&= \frac{1}{\cos\theta\sin\theta} \\
&= \sec\theta\csc\theta
\end{aligned}
$$

11. not an identity; Let $\theta = \frac{\pi}{4}$ then $\sin \theta + 1 = \sqrt{3}$ and then $\theta = 47.06°$.

13. not an identity; Let $\theta = \frac{\pi}{4}$ then $\sin^2 \frac{\pi}{4} + \cos \frac{\pi}{4} = 1.207107$.

15. $\dfrac{\tan(\alpha + \beta) - \tan \beta}{1 + \tan(\alpha + \beta) \tan \beta}$ is the formula used on $\tan((\alpha + \beta) - \beta)$ and this simplifies to $\tan \alpha$.

17.

19.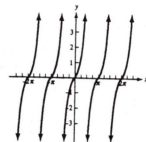

21.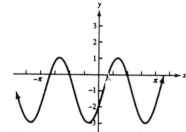

23. $b^2 = (24)^2 + (61)^2 - 2(24)(61) \cos 58$ and $b = 52$; $\dfrac{\sin 58}{52.39653} = \dfrac{\sin \alpha}{24}$ and $\alpha = 23°$; $\gamma = 180 - 58 - 23 = 99°$

25. $\alpha = 180 - 50 - 100 = 30°$; $\dfrac{\sin 30}{30} = \dfrac{\sin 50}{b}$ and $b = 46$; $\dfrac{\sin 30}{30} = \dfrac{\sin 100}{c}$ and $c = 59$

27. $\beta = 180 - 50 - 60 = 70°$; $\dfrac{\sin 70}{40} = \dfrac{\sin 50}{a}$ and $a = 33$; $\dfrac{\sin 70}{40} = \dfrac{\sin 60}{c}$ and $c = 37$

29. $c^2 = (15)^2 + (8.0)^2 - 2(15)(8.0) \cos 38$ and $c = 10$; $\dfrac{\sin \beta}{8.0} = \dfrac{\sin 38}{9.99}$ and $\beta = 30°$; $\alpha = 180 - 30 - 38 = 112°$

31. $a^2 = (14)^2 + (12)^2 - 2(14)(12) \cos 82$ and $a = 17$; $\dfrac{\sin \beta}{14} = \dfrac{\sin 82}{17.1}$ and $\beta = 54°$; $\gamma = 180 - 82 - 54 = 44°$

33. $\cos \alpha = \dfrac{(8.0)^2 + (2.0)^2 - (7.0)^2}{2(8.0)(2.0)}$ and $\alpha = 54°$; $\cos \beta = \dfrac{(7.0)^2 + (2.0)^2 - (8.0)^2}{2(7.0)(2.0)}$ and $\alpha = 113°$; $\gamma = 180 - 54 -$ $113 = 13°$

35. No solution since $a + b < c$. No triangle is formed.

37. $(14.5)^2 = (17.2)^2 + c^2 - 2(17.2)c\cos 35.5°$ and $c^2 - 34.4\cos 35.5°c + 85.59 = 0$. Then using the quadratic equation, $c = 24.5$ and $c = 3.5$. There are two triangles. When $c = 24.5$, $\dfrac{\sin\beta}{17.2} = \dfrac{\sin 35.5°}{14.5}$ and solving for β gives $\beta = 43.5°$ and $\gamma = 180 - 35.5 - 43.5 = 101.0°$. Then when $c = 3.5$, $\dfrac{\sin\gamma}{17.2} = \dfrac{\sin 35.5°}{14.5}$ and solving for γ gives $\gamma = 8.0°$. Then $\beta = 180 - 35.5 - 8.0 = 136.5°$.

39. $c^2 = (121)^2 + (315)^2 - 2(121)(315)\cos 50.0°$ and $c = 254.7 \approx 255$. Then to solve for α, $\dfrac{\sin\alpha}{121} = \dfrac{\sin 50.0°}{254.7}$ and $\alpha = 21.3°$. Then $\beta = 180 - 21.3 - 50.0 = 108.7°$.

41.
$$\cos\alpha = \frac{(9.0)^2 + (8.0)^2 - (11)^2}{2(9.0)(8.0)} \qquad \cos\beta = \frac{(11)^2 + (8.0)^2 - (9.0)^2}{2(11)(8.0)}$$

and $\alpha = 80°$ and $\beta = 54°$ so $\gamma = 180 - 80 - 54 = 46°$.

43.
$$\cos\alpha = \frac{(32.5)^2 + (35.1)^2 - (12.4)^2}{2(32.5)(35.1)} \qquad \cos\beta = \frac{(12.4)^2 + (35.1)^2 - (32.5)^2}{2(12.4)(35.1)}$$

and $\alpha = 20.7°$ and $\beta = 67.8°$ so $\gamma = 180 - 20.7 - 67.8 = 91.5°$. (The Law of Cosines will give the same answer.)

45. No solution since $\alpha + \beta > 180°$. No triangle is formed.

47. $\alpha = 180 - 15.0 - 18.0 = 147.0°$ and solving for a and c gives $\dfrac{a}{\sin 147.0°} = \dfrac{23.5}{\sin 15.0°}$ and $\dfrac{c}{\sin 18.0°} = \dfrac{23.5}{\sin 15.0°}$ so $a = 49.5$ and $c = 28.1$.

49. $(52.2)^2 = a^2 + (82.5)^2 - 2a(82.5)\cos 23.1°$ and $a^2 - 165\cos 23.1°a + 4081.41 = 0$ and using the quadratic formula gives $a = 116.8$ and $a = 34.9$. There are two triangles. Using $a = 116.8$ we find $\dfrac{\sin\beta}{82.5} = \dfrac{\sin 23.1°}{52.2}$ gives $\beta = 38.3°$. So $\alpha = 180 - 38.3 - 23.1 = 118.6°$. Then using $a = 34.9$ we find $\dfrac{\sin\alpha}{34.9} = \dfrac{\sin 23.1°}{52.2}$ gives $\alpha = 15.2°$. So $\beta = 180 - 15.2 - 23.1 = 141.7°$.

51. $(10.2)^2 = (11.8)^2 + c^2 - 2(11.8)c\cos 47.1°$ and $c^2 - 23.6\cos 47.0°c + 35.2 = 0$ and using the quadratic formula gives $c = 13.5$ and $c = 2.6$. There are two triangles. Using $c = 13.5$ we find $\dfrac{\sin\beta}{11.8} = \dfrac{\sin 47.0°}{10.2}$ gives $\beta = 57.8°$. So $\gamma = 180 - 47.0 - 57.8 = 75.2°$. Then using $c = 2.6$ we find $\dfrac{\sin\gamma}{2.6} = \dfrac{\sin 47.0°}{10.2}$ gives $\gamma = 10.8°$. So $\beta = 180 - 47.0 - 10.8 = 122.2°$.

53. Since we are given SAS and letting x be the length of the tunnel we solve for x and $x^2 = (1000)^2 + (1236)^2 - 2(1000)(1236)\cos 50.6°$ so $x = 979$ ft.

55.   a. $\sin\beta = \dfrac{\sin 20°}{1.63}$ and solving for β gives $\beta \approx 12.1°$

     b. $\sin\beta = \dfrac{\sin 35.0°}{1.63}$ and solving for β gives $\beta \approx 20.6°$

57. When the figure is constructed as indicated, we should form a right triangle with the x-axis to determine the angle at the vertex at Nut Tree. We find it to be $180° - 90° - (138° - 90°) = 52°$. Then we can use the Law of Sines to find the angle at the vertex where the time must be logged, that is, $\dfrac{\sin 52°}{200} = \dfrac{\sin \theta}{250}$ imples $\theta = 80°$. Then we can use the Law of Sines again to find the length of the trip from Nut Tree to the current position, that is. $\dfrac{200}{\sin 52°} = \dfrac{x}{\sin 48°}$ implies $x = 189$ miles. Then we can compute the time needed to fly to the current position as 1.18125 h, which would make the time be $1:10:53$ P.M.

59. Let x be the distance from the Tower 1 to the fire and y be the distance from the Tower 2 to the fire. Let γ be the included angle between x and y. $\gamma = 180° - 35.2° - 41.8° = 103°$. Then $\dfrac{\sin 103°}{10520} = \dfrac{\sin 35.2°}{x}$, and $x \approx 6224$ and $\dfrac{\sin 103°}{10520} = \dfrac{\sin 41.8°}{y}$, and $y \approx 7196$.

CHAPTER 6   COMPLEX NUMBERS AND POLAR-FORMING GRAPHING

Problem Set 6.1

1. $i = \sqrt{-1}$

3. A complex number is simplified when it is written in the form $a + bi$ where a and b are simplified real numbers.

5. Every real number is a complex, but not every complex number is a real number.

7. false; $i = \sqrt{-1}$

9. false, $(2 + \sqrt{4i})(2 - \sqrt{4i}) = (4 + 4) = 8.$

11. a. $\sqrt{-49} = \sqrt{49}i = 7i$ b. $\sqrt{-8} = \sqrt{8}i = 2\sqrt{2}i$

13. a. $\frac{-3\sqrt{-144}}{5} = \frac{-3i\sqrt{144}}{5} = \frac{-36}{5}i$ b. $\frac{-6\sqrt{-4}}{8} = \frac{-6i\sqrt{4}}{8} = \frac{-12i}{8} = \frac{-3}{2}i$

15. a. $2 + \sqrt{2} - 4 + \sqrt{-2} = -2 + \sqrt{2} + \sqrt{2}i$

    b. $6 - \sqrt{3} - 8 + \sqrt{-3} = -2 - \sqrt{3} + \sqrt{3}i$

17. a. $i(5 - 2i) = 5i - 2i^2 = 2 + 5i$

    b. $i(2 + 3i) = 2i + 3i^2 = -3 + 2i$

19. a. $(8 - 2i)(8 + 2i) = 64 - 4i^2 = 64 + 4 = 68$

    b. $(3 - 4i)(3 + 4i) = 9 - 16i^2 = 9 + 16 = 25$

21. a. $(5 - 4i) - (5 - 9i) = (5 - 5) + (-4i + 9i) = 5i$

    b. $(2 - 3i) - (4 + 5i) = (2 - 4) + (-3i - 5i) = -2 - 8i$

23. a. $-i^5 = (-1)i^2 i^2 i = (-1)(-1)(-1)i = -i$

    b. $i^7 = i^4 i^3 = (1)i^2 i = (1)(-1)i = -i$

25. a. $-i^9 = -i$

    b. $i^{10} = -1$

27. a. $(1-3i)^2 = (1-3i)(1-3i) = [(1)(1)-(-3)(-3)]+[(1)(-3)+(-3)(1)]i = 1-9+(-3-3)i = -8 - 6i$

    b. $(6 - 2i)^2 = (6 - 2i)(6 - 2i) = [(6)(6) - (-2)(-2)] + [(6)(-2) + (-2)(6)]i = (36 - 4) + (-12 - 12)i = 32 - 24i$

29. a. $(\sqrt{2}+3i)^2 = (\sqrt{2}+3i)(\sqrt{2}+3i) = 2+\sqrt{2}3i+\sqrt{2}3i+9i^2 = 2+6\sqrt{2}i+(-9) = -7+6\sqrt{2}i$

    b. $(\sqrt{5} - 3i)^2 = (\sqrt{5} - 3i)(\sqrt{5} - 3i) = 5 - 3\sqrt{5}i - 3\sqrt{5}i + 9i^2 = 5 - 6\sqrt{5}i - 9 = -4 - 6\sqrt{5}i$

31. a. $(3 - \sqrt{-3})(3 + \sqrt{-3}) = 9 - (\sqrt{-3})^2 - 3\sqrt{-3} + 3\sqrt{-3} = 9 - (-3) = 12$

    b. $(2 - \sqrt{-3})(2 + \sqrt{-3}) = 4 - (\sqrt{-3})^2 + 2\sqrt{-3} - 2\sqrt{-3} = 4 - (-3) = 7$

33. $\frac{4-2i}{3+i} = \frac{[(4)(3)+(-2)(1)]}{(3^2+1^2)} + \frac{[(-2)(3)-(4)(1)]}{(3^2+1^2)}i = \frac{12-2}{10} + \frac{-6-4}{10}i = 1 - i$

35. $\frac{1+3i}{1-2i} = \frac{[(1)(1)+(3)(-2)]}{(1^2+(-2)^2)} + \frac{[(3)(1)-(1)(-2)]}{(1^2+(-2)^2)}i = \frac{1-6}{5} + \frac{3+2}{5}i = -1 + i$

37. $\frac{-3}{1+i} = \frac{[(-3)(1)+(1)(0)]}{1^2+1^2} + \frac{[(0)(1)-(-3)(1)]}{1^2+1^2}i = \frac{-3}{2} + \frac{3}{2}i$

39. $\frac{2}{i} = \frac{2}{i}\frac{i}{i} = \frac{2i}{-1} = -2i$

41. $\frac{-2i}{3+i} = \frac{[(0)(3)+(-2)(1)]}{3^2+1^2} + \frac{[(-2)(3)-(0)(1)]}{3^2+1^2}i = -\frac{1}{5} - \frac{3}{5}i$

43. $\frac{-i}{2-i} = \frac{[(0)(2)+(-1)(-1)]}{2^2+(-1)^2} + \frac{[(-1)(2)-(0)(-1)]}{2^2+(-1)^2}i = \frac{1}{5} - \frac{2}{5}i$

45. $\frac{2+71}{2-7i} = \frac{[(2)(2)+(-7)(7)]}{2^2+(-7)^2} + \frac{[(7)(2)-(2)(-7)]}{2^2+(-7)^2} = -\frac{45}{53} + \frac{28}{53}i$

47. $\frac{\sqrt{-1}+1}{\sqrt{-1}-1} = \frac{i+1}{i-1} = \frac{[(1)(-1)+(1)(1)]}{(-1)^2+(1)^2} + \frac{[(1)(-1)-(1)(1)]}{(-1)^2+1^2}i = \frac{-1-1}{2}i = -i$

49. $\frac{4-2i}{1+i} = \frac{[(4)(1)+(-2)(1)]}{1^2+1^2} + \frac{[(-2)(1)-(4)(1)]}{1^2+1^2}i = \frac{4-2}{2} + \frac{-2-4}{2}i = 1 - 3i$

51. $x^2 + 25, x = 5i$ $(5i)^2 + 25 = 25(i)^2 + 25 = -25 + 25 = 0$

53. $x^2 + 2 = 0$ so $x^2 = -2$ and $x = \pm\sqrt{-2}$ or $x = \pm\sqrt{2}i$

55. $x^2 - 2x + 5$, so $x = 1 + 2i$ then $(1+2i)^2 - 2(1+2i) + 5 = 1 + 4i + 4(-1) - 4i + 3 = 0$.

57. Use the quadratic formula, where $b = -3, a = 1, c = 8$. Thus, $x = \frac{3\pm\sqrt{23}}{2}i$.

59. $x^3 - 11x^2 + 40x - 50$ and $x = 3 + 1$ then $(3+i)^3 - 11(3+i)^2 + 40(3+i) - 50 = (3+i)(8 + 6i) - 11(8 + 6i) + 70 + 40i = 0$.

61. $x^3 - 1 = 0$ and $(x-1)(x^2+x+1) = 0$. Factoring and using the quadratic formula, $x = 1$ and $x = \frac{-1\pm\sqrt{3}}{2}i$.

63. $(1.9319 + 0.5176i)(2.5985 + 1.5i) = 4.2429 + 4.2426i$

65.

$$
\begin{aligned}
z_1 + z_2 &= (a + bi) + (c + di) && \text{Substitution} \\
&= (a + c) + (b + d)i && \text{Definition of addition} \\
&= (c + a) + (d + b)i && \text{Communtative law for real numbers} \\
&= (c + di) + (a + bi) && \text{Defintion of addition} \\
&= z_2 + z_1 && \text{Substitution} \\
z_1 z_2 &= (a + bi)(c + di) && \text{Substitution} \\
&= (ac - bd) + (ad + bc)i && \text{Defintion of multiplication} \\
&= (ca - db) + (da + cb)i && \text{Communtative law for real numbers} \\
&= (c + di)(a + bi) && \text{Definition of multiplication} \\
&= z_2 z_1 && \text{Substitution}
\end{aligned}
$$

Problem Set 6.2

9. false; $\theta = 0°$ or $180°$

11. false; It is plotted on the positive imaginary axis.

13. false; 4cis 250° is in Quadrant III

15.   a. $|z| = \sqrt{(-3)^2 + (-2)^2} = \sqrt{13}$ b. $|z| = \sqrt{2^2 + 4^2} = \sqrt{20}$
        c. $|z| = \sqrt{5^2 + 6^2} = \sqrt{61}$

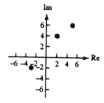

17.   a. $|z| = \sqrt{2^2 + (-5)^2} = \sqrt{29}$ b. $|z| = \sqrt{4^2 + (-1)^2} = \sqrt{17}$
        c. $|z| = \sqrt{(-1)^2 + 1^2} = \sqrt{2}$

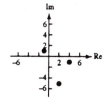

19. a. $r = \sqrt{(\sqrt{3})^2 + 1^2} = \sqrt{3+1} = \sqrt{4} = 2$ and $\tan\theta = \frac{b}{a} = \frac{1}{\sqrt{3}}$ so $\theta = 30°$ and $\sqrt{3} + i = $ 2cis 30°.

   b. $r = \sqrt{(1^2 + (-\sqrt{3})^2} = \sqrt{1+3} = \sqrt{4} = 2$ and $\tan\theta = \frac{b}{a} = \frac{-\sqrt{3}}{1}$ so $\theta = 300°$ and $1 - \sqrt{3}i = $ 2cis 300°.

   c. $r = \sqrt{(-1)^2 + (-\sqrt{3})^2} = \sqrt{1+3} = \sqrt{4} = 2$ and $\tan\theta = \frac{b}{a} = \frac{-\sqrt{3}}{-1}$ so $\theta = 240°$ and $-1 - \sqrt{3}i = $ 2cis 240°.

21. a. $r = \sqrt{0^2 + 1^2} = 1$ and $\tan\theta = \frac{1}{0}$ is undefined, so $\theta = 90°$ cis 90°.

   b. $r = \sqrt{0^2 + (-2)^2} = \sqrt{4} = 2$ and $\tan\theta = \frac{-2}{0}$ is undefined, so $\theta = 270°$ 2cis 270°.

   c. $r = \sqrt{0^2 + 3^2} = \sqrt{9} = 3$ and $\tan\theta = \frac{3}{0}$ is undefined, so $\theta = 90°$ 3cis 90°.

23. a. $r = \sqrt{(-0.6946)^2 + (3.9392)^2} = 3.9999707 \approx 4$ and $\tan\theta = \frac{3.9392}{-0.6946}$ so $\theta = 100°$ and 4cis 100°.

   b. $r = \sqrt{(-2.0337)^2 + (-4.5677)^2} = 5$ and $\tan\theta = \frac{-4.5677}{-2.0337}$ so $\theta = 246°$ and 5cis 246°.

25. a. $3(\cos 60° + i\sin 60°) = 3\left(\frac{1}{2} + \frac{\sqrt{3}}{2}i\right) = \frac{3}{2} + \frac{3\sqrt{3}}{2}i$

   b. $5(\cos\frac{3\pi}{2} + i\sin\frac{3\pi}{2}) = 5(0 - i) = -5i$

27. a. $2.5(\cos 300° + i\sin 300°) = 1.25 - \frac{2.5\sqrt{3}}{2}i$ or $\frac{5}{4} - \frac{5\sqrt{3}}{2}i$

   b. $4.2(\cos 135° + i\sin 135°) = 4.2(-\frac{\sqrt{2}}{2} + \frac{\sqrt{2}}{2}i) = -2.1\sqrt{2} + 2.1\sqrt{2}i$

29. $(3\text{cis } 48°)(5\text{cis } 92°) = (3)(5)\text{cis } (48° + 92°) = 15\text{cis } 140°$

31.

$$\frac{5(\cos 315° + i\sin 315°)}{2(\cos 48° + i\sin 48°)} = \frac{5\text{cis } 315°}{2\text{cis } 48°} = \frac{5}{2}\text{cis } 267°$$

33.

$$\frac{12\text{cis } 250°}{4\text{cis } 120°} = \frac{12}{4}\text{cis } (250° - 120°) = 3\text{cis } 130°$$

35. $(3\text{cis } 60°)^4 = 3^4\text{cis } (4(60°)) = 81\text{cis } 240°$

37. $(2\text{cis } 80°)^6 = 2^6\text{cis } (6(80°)) = 64\text{cis } 480° = 64\text{cis } 120°$

39. $r = \sqrt{(\sqrt{3})^2 + (-1)^2} = \sqrt{4} = 2$ and $\tan\theta = \frac{-1}{\sqrt{3}}$ so $\theta = 330°$ and $(2\text{cis } 330°)^8 = 2^8\text{cis } (8(330°)) = 256\text{cis } 120°$.

41. $(81)^{1/4}\text{cis } \left(\frac{88° + 360k°}{4}\right) = 3\text{cis } (22° + 90k°)$. So, for $k = 0$; 3cis 22°; for $k = 1$; 3cis 112°; for $k = 2$; 3cis 202°; for $k = 3$; 3cis 292°.

43. $27 = 27\text{cis } 0°$ so $(27\text{cis } 0°)^{1/3} = 27^{1/3}\text{cis } \left(\frac{0°+360k°}{3}\right) = 3\text{cis } (0+120k)$. So, for $k = 0$, $3\text{cis } 0°$; for $k = 1$, $3\text{cis } 120°$; for $k = 2$, $3\text{cis } 240°$.

45. $r = \sqrt{1^2 + 1^2} = \sqrt{2}$ and $\tan\theta = 1$, so $\theta = 45°$ then $(\sqrt{2}\text{cis } 45°)^{1/4} = \sqrt[8]{2}\text{cis } \left(\frac{45°+360k°}{4}\right)$. So, for $k = 0$, $\sqrt[8]{2}\text{cis } 11.25°$; for $k = 1$, $\sqrt[8]{2}\text{cis } 101.25°$; for $k = 2$, $\sqrt[8]{2}\text{cis } 191.25°$; for $k = 3$, $\sqrt[8]{2}\text{cis } 281.25°$.

47. $(32\text{cis } 200°)^{1/5} = 32^{1/5}\text{cis } \left(\frac{200°+360k°}{5}\right) = 2\text{cis } (40° + 72k°)$. So, for $k = 0$, $2\text{cis } 40°$; for $k = 1$, $2\text{cis } 112°$; for $k = 2$, $2\text{cis } 184°$; for $k = 3$, $2\text{cis } 328°$.

49. $(-8\text{cis } 180°)^{1/3} = -2\text{cis } \left(\frac{180°+360k°}{3}\right) = -2\text{cis } (60° + 120k°)$. So, for $k = 0$, $-2\text{cis } 60°$ or $1 + \sqrt{3}i$; for $k = 1$, $-2\text{cis } 180°$ or -2; for $k = 2$, $-2\text{cis } 300°$ or $1 - \sqrt{3}i$.

51. $r = \sqrt{(4\sqrt{3})^2 + (-4)^2} = \sqrt{48 + 16} = \sqrt{64} = 8$ and $\tan\theta = \frac{-4}{4\sqrt{3}} = \frac{-1}{\sqrt{3}}$ so $\theta = 330°$ and $(8\text{cis } 330°)^{\frac{1}{3}} = 8^{1/3}\text{cis } \left(\frac{330°+360k°}{3}\right) = 2\text{cis } (110° + 120k°)$. So, for $k = 0$, $2\text{cis } 110°$ or $-0.6840 + 1.8794i$; for $k = 1$, $2\text{cis } 230°$ or $-1.2858 - 1.5321i$; for $k = 2$, $2\text{cis } 350°$ or $1.9896 - 0.3473i$.

53. $r = \sqrt{(12.2567)^2 + (10.2846)^2} = 16$ and $\tan\theta = \frac{10.2846}{12.2567}$ so $\theta = 40°$ and $(16\text{cis } 40°)^{\frac{1}{4}} = 16^{1/4}\text{cis } \left(\frac{40°+360k°}{4}\right) = 2\text{cis } (10° + 90k°)$. So, for $k = 0$, $2\text{cis } 10°$ or $1.9696 + 0.3473i$; for $k = 1$, $2\text{cis } 100°$ or $-0.3473 + 1.9696i$; for $k = 2$, $2\text{cis } 190°$ or $-1.9696 - 0.3473i$; for $k = 3$, $2\text{cis } 280°$ or $0.3473 - 1.9696i$.

55. $r = 1$ and $\theta = 0°$ so $(1\text{cis } 0°)^{1/5} = 1^{1/5}\text{cis } \left(\frac{0°+360k°}{5}\right) = 1\text{cis } (0° + 72k°)$. So, for $k = 0$, $\text{cis } 0°$ or 1 for $k = 1$, $\text{cis } 72°$ or $-0.3090 + 0.9511i$; for $k = 2$, $\text{cis } 144°$ or $-0.8090 + 0.5878i$; for $k = 3$, $\text{cis } 216°$ or $-0.8090 - 0.5878i$; for $k = 4$, $\text{cis } 288°$ or $0.3090 - 0.9511i$.

57. $r = 1$ and $\theta = 0°$ so $(1\text{cis } 0°)^{1/6} = 1^{1/6}\text{cis } \left(\frac{0°+360k°}{6}\right) = 1\text{cis } (0° + 60k°)$. So, for $k = 0$, $\text{cis } 0°$ or 1; for $k = 1$, $\text{cis } 60°$ or $\frac{1}{2} + \frac{\sqrt{3}}{2}i$; for $k = 2$, $\text{cis } 120°$ or $-\frac{1}{2} + \frac{\sqrt{3}}{2}i$; for $k = 3$, $\text{cis } 180°$ or -1; for $k = 4$, $\text{cis } 240°$ or $-\frac{1}{2} - \frac{\sqrt{3}}{2}i$; for $k = 5$, $\text{cis } 320°$ or $\frac{1}{2} - \frac{\sqrt{3}}{2}i$.

59. $r = 16$ and $\theta = 0°$ and $(16\text{cis } 0°)^{1/4} = 16^{1/4}\text{cis } \left(\frac{0°+360k°}{4}\right) = 16^{1/4}\text{cis } (0° + 90k°)$. So, for $k = 0$, $\text{cis } 0°$ or 2; for $k = 1$, $\text{cis } 90°$ or $2i$; for $k = 2$, $\text{cis } 180°$ or -2; for $k = 3$, $\text{cis } 270°$ or $-2i$.

61. By multiplying the equation by $(x - 1)$, gives $r = 1$ and $\theta = 0°$ so $(1\text{cis } 0°)^{1/6} = 1^{1/6}\text{cis } \left(\frac{0°+360k°}{6}\right) = 1\text{cis } (0° + 60k°)$. So, for $k = 0$, the function is zero and not relevant; for $k = 1$, $\text{cis } 60°$ or $\frac{1}{2} + \frac{\sqrt{3}}{2}i$; for $k = 2$, $\text{cis } 120°$ or $-\frac{1}{2} + \frac{\sqrt{3}}{2}i$; for $k = 3$, $\text{cis } 180°$ or -1; for $k = 4$, $\text{cis } 240°$ or $-\frac{1}{2} - \frac{\sqrt{3}}{2}i$; for $k = 5$, $\text{cis } 320°$ or $\frac{1}{2} - \frac{\sqrt{3}}{2}i$.

63.

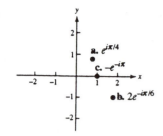

65.

$$I = \frac{E}{Z} = \frac{12\text{cis } 27°}{\sqrt{13}\text{cis } 33.69°}$$

$$= \frac{12}{\sqrt{13}}\text{cis } (27° - 33.69°)$$

$$= \frac{12}{\sqrt{13}}(\cos(-6.69°) + i\sin(-6.69°))$$

$$\approx 3.31 - 0.39i$$

67. $\mathbf{F}_1 = \langle 25\cos 30°, 25\sin 30° \rangle$ and $\mathbf{F}_2 = \langle 45\cos 60°, 45\sin 60° \rangle$. Then $\mathbf{F}_1 + \mathbf{F}_2 = \langle 25\left(\frac{\sqrt{3}}{2}\right) + 45\left(\frac{1}{2}\right), 25\left(\frac{1}{2}\right) + 45\left(\frac{\sqrt{3}}{2}\right) \rangle$. Then solving for r in $r\text{cis}\theta$ gives

$$r = \sqrt{\left(25\left(\frac{\sqrt{3}}{2}\right) + 45\left(\frac{1}{2}\right)\right)^2 + \left(25\left(\frac{1}{2}\right) + 45\left(\frac{\sqrt{3}}{2}\right)\right)^2} \approx 68$$

and solving for θ gives $\theta \approx 49°$. So the equation is $68\text{cis } 49°$.

Problem Set 6.3

5. polar: $\left(4, \frac{\pi}{4}\right)$, $\left(-4, \frac{5\pi}{4}\right)$; $x = 4\cos\frac{\pi}{4}$ and $y = 4\sin\frac{\pi}{4}$; rectangular: $(2\sqrt{2}, 2\sqrt{2})$

7. polar: $\left(5, \frac{2\pi}{3}\right)$, $\left(-5, \frac{5\pi}{3}\right)$; $x = 5\cos\frac{2\pi}{3}$ and $y = 5\sin\frac{2\pi}{3}$; rectangular: $\left(-\frac{5}{2}, \frac{5\sqrt{3}}{2}\right)$

9. polar: $\left(\frac{3}{2}, \frac{7\pi}{6}\right)$, $\left(-\frac{3}{2}, \frac{\pi}{6}\right)$; $x = \frac{3}{2}\cos\frac{7\pi}{6}$ and $y = -\frac{3}{2}\sin\frac{\pi}{6}$; rectangular: $\left(\frac{3\sqrt{3}}{4}, -\frac{3}{4}\right)$

11. polar: $(1, \pi), (-1, 0)$; $x = \cos\pi$ and $y = \sin\pi$; rectangular: $(-1, 0)$

13. polar: $(0, 2\pi - 3)$, $(0, \pi - 3)$; $x = 0\cos(2\pi - 3)$ and $y = 0\sin(\pi - 3)$; rectangular: $(0, 0)$

15. $r = \sqrt{(-1)^2 + (\sqrt{3})^2} = \sqrt{4} = 2$ and $\left(\tan \frac{\sqrt{3}}{-1}\right)^{-1} = (\tan - \sqrt{3})^{-1} = \theta$. So $\theta = \frac{-\pi}{3}$ and $\theta = \frac{2\pi}{3}$ then $\left(2, \frac{2\pi}{3}\right)$ and $\left(-2, \frac{5\pi}{3}\right)$

17. $r = \sqrt{(-2)^2 + (-2)^2} = \sqrt{8}$ and $(\tan - 1)^{-1} = \theta$. So $\theta = -\frac{\pi}{4}$ and $\theta = \frac{5\pi}{4}$ then $\left(\sqrt{8}, \frac{5\pi}{4}\right)$ and $\left(-\sqrt{8}, \frac{\pi}{4}\right)$

19. $r = \sqrt{3^2 + 7^2} = 7.62$ and $\left(\tan \frac{7}{3}\right)^{-1} = \theta$, so $\theta = 1.17$ and $\theta = 4.31$ then $(7.62, 1.17)$ and $(-7.62, 4.31)$

21. $r = \sqrt{(\sqrt{3})^2 + (-1)^2} = 2$ and $\left(\tan \frac{1}{\sqrt{3}}\right)^{-1} = \theta$, so $\theta = -\frac{\pi}{6}$ and $\theta = \frac{5\pi}{6}$ then $\left(2, \frac{5\pi}{6}\right)$ and $\left(-2, \frac{11}{6}\right)$

23. $r = 4\sin\theta$ and $r^2 = 4r\sin\theta$ so $x^2 + y^2 = 4y$ or $x^2 + y^2 - 4y = 0$

25. $r = \sec\theta = \frac{1}{\cos\theta}$ and $\frac{r}{r} = \frac{1}{r\cos\theta}$ so $1 = \frac{1}{x}$ and $x = 1$

27. $r^2 + r^2\sin^2\theta = 2$ and $x^2 + y^2 + y^2 = 2$ thus $x^2 + 2y^2 = 2$

29. $r = 1 - \sin\theta$ and $r^2 = r - r\sin\theta$ so $x^2 + y^2 = r - y$ or $x^2 + y^2 = \sqrt{x^2 + y^2} - y$

31.

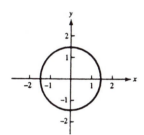

33.

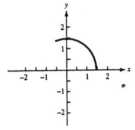

35.

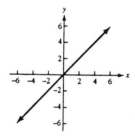

37.

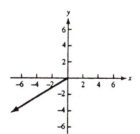

39. yes   **41.** no   **43.** yes   **45.** yes

47. No, because $-1 = 2\left(1 - \cos\frac{\pi}{3}\right)$ is not true, $-1 \neq 1$. Also, we must check $\left(1, \frac{4\pi}{3}\right)$, but this doesn't lie on the curve since $1 = 2\left(1 - \cos\frac{4\pi}{3}\right)$ is false.

49. No, since $2 + \sqrt{2} = 2\left(1 - \cos\frac{\pi}{4}\right)$ is not true. Also, we check $\left(-2 - \sqrt{2}, \frac{5\pi}{4}\right)$. This is false also, since $-2 - \sqrt{2} \neq 2\left(1 - \cos\frac{5\pi}{4}\right)$.

51. No, because $0 \neq 2\left(1 - \cos\frac{\pi}{4}\right)$. We check $\left(0, \frac{5\pi}{4}\right)$ also, but it is false too, since $0 \neq 2\left(1 - \cos\frac{5\pi}{4}\right)$.

53. Let $\theta = \frac{\pi}{2}$. Then $r^2 = 9\cos\theta = 0$, so the coordinates are $\left(0, \frac{\pi}{2}\right) = \left(0, \frac{3\pi}{2}\right)$. Let $\theta = 0$. Then $r^2 = 9$ and $r = \pm 3$. So the coordinates are $(3, 0) = (-3, \pi)$ and $(-3, 0) = (3, \pi)$.

55. Let $\theta = \pi$. Then $r = 3\pi$ and the coordinates are $(3\pi, \pi) = (-3\pi, 2\pi)$. Let $\theta = \frac{\pi}{2}$. Then $r = \frac{3\pi}{2}$ and the coordinates are $\left(\frac{3\pi}{2}, \frac{\pi}{2}\right) = \left(-\frac{3\pi}{2}, \frac{3\pi}{2}\right)$. Let $\theta = \frac{\pi}{4}$. Then $r = \frac{3\pi}{4}$. The coordinates are $\left(\frac{3\pi}{4}, \frac{\pi}{4}\right) = \left(-\frac{3\pi}{4}, \frac{\pi}{4}\right)$.

57. Let $\theta = 0$. Then $r = 2 - 3\sin\theta = 2$ and the coordinates are $(2, 0) = (-2, \pi)$. Let $\theta = \pi$. Then $r = 2 - 3\sin\theta = -1$ and the coordinates are $(-1, \pi) = (1, 2\pi)$. Let $\theta = \frac{3\pi}{2}$. Then $r = 5$. The coordinates are $\left(5, \frac{3\pi}{2}\right) = \left(-5, \frac{5\pi}{2}\right)$.

59. Let $\theta = 0$. Then $r = 2 - 2\sin\theta = 2$ and the coordinates are $(2, 0) = (-2, 0)$. Let $\theta = \pi$. Then $r = 2 - 2\sin\theta = 0$ and the coordinates are $(0, \pi) = (0, 2\pi)$. Let $\theta = \frac{3\pi}{2}$. Then $r = 4$. The coordinates are $\left(4, \frac{3\pi}{2}\right) = \left(-4, \frac{5\pi}{2}\right)$.

61.   a.   United States   b.   India   c.   Greenland   d. Canada

63.

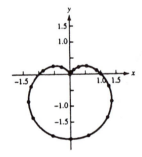

65.

$$\left\{(r, \theta) \,\middle|\, a^2 = (r\cos\theta - R\cos\alpha)^2 + (r\sin\theta - R\sin\alpha)^2\right\}$$

$$= r^2\cos^2\theta - 2rR\cos\theta\cos\alpha + R^2\cos^2\alpha + r^2\sin^2\theta - 2rR\sin\theta\sin\alpha + R^2\sin^2\alpha$$

$$= r^2 + R^2 - 2rR(\cos\theta\cos\alpha + \sin\theta\sin\alpha)$$

$$= r^2 + R^2 - 2rR\cos(\theta - \alpha)$$

Problem Set 6.4

5. a. four-leaved rose b. lemniscate c. circle d. sixteen-leaved rise

e. none f. lemniscate g. three-leaved rose h. cardioid

7.

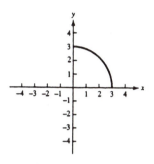

9.

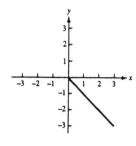

11.

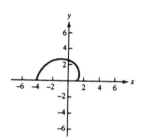

13.

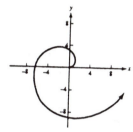

15.

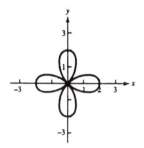

17.

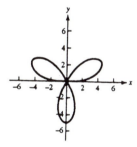

19.

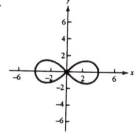

21.

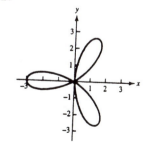

23.

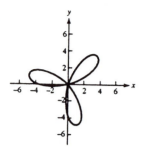

25.

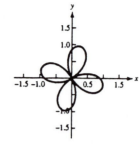

27.

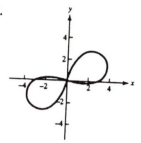

29.

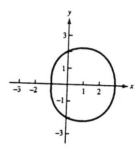

31.

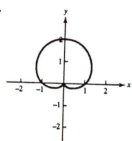

33.

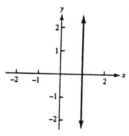

35.

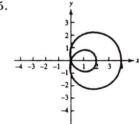

37.

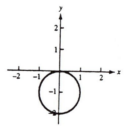

39.

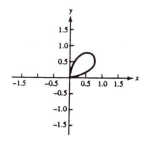

41.

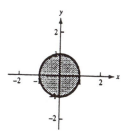

43.

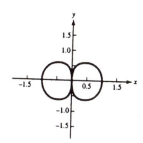

45.

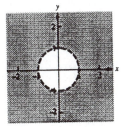

47.

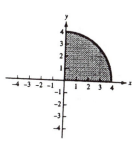

49.

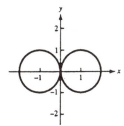

51.

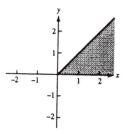

53.

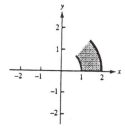

55.
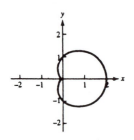

57.
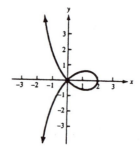

59.
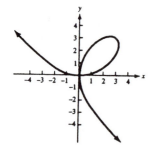

61.
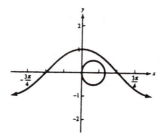

63.
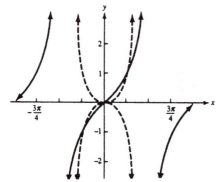

65.
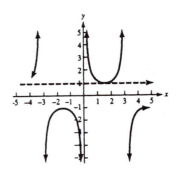

69. If the equation is unaffected when r is replaced by $-r$ and θ by $\pi - \theta$, then the points (r, θ) and $(-r, \pi - \theta)$ are both on the graph. This implies symmetry with respect to the x-axis.

71. If the equation is unaffected when r is replaced by $-r$ and θ by $\frac{\pi}{2} - \theta$, then the points (r, θ) and $\left(-r, \frac{\pi}{2} - \theta\right)$ are both on the graph. This implies symmetry with respect to the origin.

73.   a.   This follows immediately from the fact that (r, θ) is on the original graph if and only if $(r, \theta - \alpha)$ is on the new graph.

      b.   The graph of $r = 2 \sec \theta$ is $r \cos \theta = 2$ or $x = 2$. Then rotate through $\theta = \frac{\pi}{3}$.

Chapter 6 Sample Test

1. a. $2 + 5i$ b. $-1 - 3i$ c. 8

     d. $-21 - 20i$ e. $-\frac{1}{2} + \frac{5}{2}i$

3. a. $r = \sqrt{7^2 + 7^2} = \sqrt{98} = 7\sqrt{2}$ and $\theta = 315°$ which gives $7\sqrt{2}\text{cis } 315°$.

     b. Since $-3i$ is on the negative imaginary axis, $\theta = 270°$ and $r = \sqrt{3^2} = 3$ which gives $3\text{cis } 270°$.

     c. $2\text{cis } 150° = 2(\cos 150° + i \sin 150° = -\sqrt{3} + i$

     d. $2\sqrt{2} - 2\sqrt{2}i$

5. $(27\text{cis } 0°)^{1/3} = \sqrt[3]{27}\text{cis } \frac{0° + 360k°}{3} = 3\text{cis } 0° + 120k°$. So, for $k = 0$, $3\text{cis } 0°$; for $k = 1$, $3\text{cis } 120°$; for $k = 2$, $3\text{cis } 240°$.

7. a. no   b. yes   c. yes   d. no

9. a.

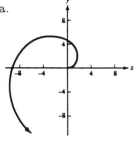

     b.

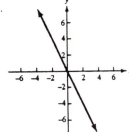

     c.

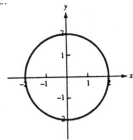

     d.

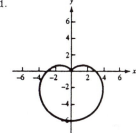

Chapter 6 Miscellaneous Problems

1. a. 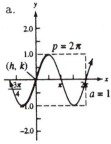   b.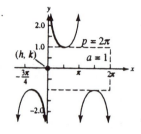

3. $y = \cos \theta$

5.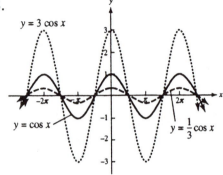

7. $y = 2 \sin \pi x$

9. $y = \frac{1}{4} \cos \frac{\pi}{3} x$

11. $1 + (\sin \theta)^2 = 1 + (1 - (\cos \theta)^2) = 2 - (\cos \theta)^2$

13.
$$\frac{(\sec^2 \theta + \tan^2 \theta)^2}{\sec^4 \theta + \tan^4 \theta} = \frac{(\sec^2 \theta + \tan^2 \theta)^2}{(\sec^2 \theta + \tan^2 \theta)(\sec^2 \theta - \tan^2 \theta)}$$
$$= \frac{\sec^2 \theta + \tan^2 \theta}{1}$$
$$= 1 + \tan^2 \theta + \tan^2 \theta$$
$$= 1 + 2 \tan^2 \theta$$

15. $\sec(-\theta) = \frac{1}{\cos(-\theta)} = \frac{1}{\cos \theta} = \sec \theta$

17. $\cos^2 x - 2\sin x + 3 = 0$ and $1 - \sin^2 x - 2\sin x + 3 = 0$ then $(-\sin x + 1)(\sin x + 4) = 0$ so $\sin x = 1$. Since $\sin x \neq -4$ and $x = \frac{\pi}{2} \approx 1.57$.

19. $2\cos 5x = 1$ and $5x = \frac{\pi}{3}$ so $x = \frac{\pi}{15} \approx 0.21$.

21. $r = \sqrt{\left(-\frac{7\sqrt{3}}{2}\right)^2 + \left(\frac{7}{2}\right)^2} = 7$ and $\tan\theta = \frac{\frac{7}{2}}{-\frac{7\sqrt{3}}{2}} = -\frac{1}{\sqrt{3}}$ so $\theta = \frac{5\pi}{6} = 150°$ which gives $7\text{cis}\,150°$.

23. $r = 16$ and $16\text{cis}\,90°$ since $16i$ lies on the positive imaginary axis.

25. $r = \sqrt{(-2)^2 + 3^2} = \sqrt{13}$ and $\tan\theta = \frac{3}{-2}$ so $\theta \approx 124°$ and $\sqrt{13}\text{cis}\,124°$.

27. $9\left(\cos\frac{3\pi}{4} + i\sin\frac{3\pi}{4}\right) = -\frac{9\sqrt{2}}{2} + \frac{9\sqrt{2}}{2}i$

29. $3(\cos 240° + i\sin 240°) = -\frac{3}{2} - \frac{3\sqrt{3}}{2}i$

31. $5(\cos 25° + i\sin 25°) = 4.5315 + 2.1131i$

33. $r = \sqrt{(\sqrt{3})^2 + (-1)^2} = \sqrt{4} = 2$ and $\tan\theta = -\frac{1}{\sqrt{3}}$ so $\theta = \frac{11\pi}{6}$ and $\left(2\text{cis}\,\frac{11\pi}{6}\right)^6 = 2^6\text{cis}\left(6\left(\frac{11\pi}{6}\right)\right)$ then $64\text{cis}\,11\pi = -64$.

35. $\frac{2\text{cis}\,185°(4\text{cis}\,223°)}{(2\text{cis}\,300°)^3} = \frac{8\text{cis}\,408°}{8\text{cis}\,900°} = \text{cis}\,-492° = \text{cis}\,228°$

37. $(\text{cis}\,90°)^{1/3} = \text{cis}\,\frac{90° + 360k°}{3} = \text{cis}\,30° + 120k°$. So, for $k = 0$, $\text{cis}\,30°$; for $k = 1$, $\text{cis}\,150°$; for $k = 2$, $\text{cis}\,270°$.

39. $1 = 1\text{cis}\,0°$ since 1 lies on the positive real axis and $(\text{cis}\,0°)^{1/10} = \text{cis}\,\frac{0° + 360k°}{10}$. So, for $k = 0$, $\text{cis}\,0°$; for $k = 1$, $\text{cis}\,36°$; for $k = 2$, $\text{cis}\,72°$; for $k = 3$, $\text{cis}\,108°$; for $k = 4$, $\text{cis}\,144°$; for $k = 5$, $\text{cis}\,180°$; for $k = 6$, $\text{cis}\,216°$; for $k = 7$, $\text{cis}\,252°$; for $k = 8$, $\text{cis}\,288°$; for $k = 9$, $\text{cis}\,324°$.

41. $r = \sqrt{(-16)^2 + (16\sqrt{3})^2} = 32$ and $\tan\theta = \frac{16\sqrt{3}}{-16} = -\sqrt{3}$ so $\theta = 120°$ and $(32\text{cis}\,120°)^{1/3} = \sqrt[3]{32}\text{cis}\,(40° + 120k°)$. So, for $k = 0$, $\text{cis}\,40°$; for $k = 1$, $\text{cis}\,160°$; for $k = 2$, $\text{cis}\,280°$.

43.

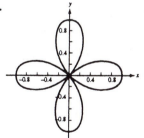

45.

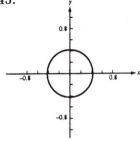

47.

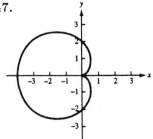

49.

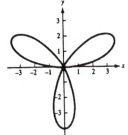

51.

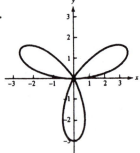

53.

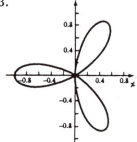

55.

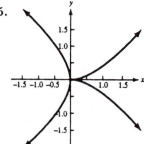

57.

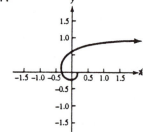

59.

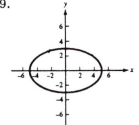

61.

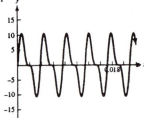

63. We assume that $t = 0$ at noon and that the balloon is released on the ground. So at time t, we have $h = 4t$ m/s. So $d(t) = \sqrt{(4t)^2 + (100)^2}$.

Chapters 4–6 Cumulative Review

23. $K = \frac{1}{2}bc \sin \alpha = \frac{1}{2}ac \sin \beta = \frac{1}{2}ab \sin \gamma$

25. $K = \sqrt{s(s-a)(s-b)(s-c)}$ where $s = \frac{1}{2}(a+b+c)$

27. The volume of a circular cone with base radius r and height h is $V = \frac{1}{3}r^2 h$. If the angle of elevation from the base to the vertex is α, then $V = \frac{1}{3}r^3 \pi \tan \alpha$.

29. $\mathbf{v} \cdot \mathbf{w} = 0$

31. $\mathbf{x} = \mathbf{r} \cos \theta$, $\mathbf{y} = \mathbf{r} \sin \theta$

33. $\theta = $ constant

35. $\mathbf{r} = \mathbf{b} \pm \mathbf{a} \cos \theta$ or $\mathbf{r} = \mathbf{b} \pm a \sin \theta$

37. $r^2 = a^2 \cos 2\theta$ or $r^2 \sin 2\theta$

39. A **41.** E **43.** E **45.** C **47.** B **49.** C **51.** B **53.** C **55.** B **57.** C

59. a. rose curve b. lemniscate c. rose curve d. cardioid

61. a. circle b. limacon c. cardioid d. line

63. $\theta = 0, \frac{\pi}{3}, \frac{2\pi}{3}, \pi, \frac{4\pi}{3}, \frac{5\pi}{3}, 2\pi$

65. $\frac{\sin 3\theta + \cos 3\theta + 1}{\cos 3\theta} = \frac{\sin 3\theta}{\cos 3\theta} \frac{1}{\cos 3\theta} + 1 = \tan 3\theta + \sec 3\theta + 1$

67. Law of Sines; $\frac{\sin \alpha}{38} = \frac{\sin 108°}{42}, \alpha = 59.37°$ $\gamma = 180° - \alpha - \beta = 12.63°$ $\frac{c}{\sin 12.63°} = \frac{42}{\sin 108°}, c = 9.656$

69. Law of Sines; $\beta = 180° - \alpha - \gamma = 68°$ $\frac{\sin 68°}{52} = \frac{\sin 49°}{a}, a = 42.33$ $\frac{\sin 68°}{52} = \frac{\sin 63°}{c}, c = 49.97$

71. Since $\beta \geq 180$, a triangle is not formed.

73. Right triangle solution; $\beta = 90°, \gamma = 60°$ $c^2 = b^2 - a^2 = 36 - 9 = 27, c = \sqrt{27}$

75. Law of Sines; $\frac{\sin 57°}{55} = \frac{\sin \beta}{61}, \beta = 68.46°$ $\gamma = 180 - \alpha - \beta = 54.54°$ $\frac{\sin 57°}{55} = \frac{\sin 54.54°}{c}, c = 53.4$

77. Primary representation $(5, \frac{5\pi}{6})$. In rectangular, $x = -5 \cos -\frac{\pi}{6}, y = -5 \sin -\frac{\pi}{6}$ so $x = -4.33, y = \frac{5}{2}$. Then, $\left(\frac{-5}{2}\sqrt{3}, \frac{5}{2}\right)$.

79. Primary representation $(3, 6.1424)$. In rectangular, $x = 3 \cos 6.1424, y = 3 \sin 6.1424$ so $x \approx 3, y \approx -0.42$. Then, $(3, -0.42)$.

81. $-8i$ lies on the negative imaginary axis, thus $\theta = 270°$. Thus, in trigonometric form, we have $8 \operatorname{cis} 270°$. For the square roots we receive $\sqrt{8} \operatorname{cis} \left(\frac{270 + 360k}{2}\right)$. So for $k = 0, \sqrt{8} \operatorname{cis} 135°$; $k = 1, \sqrt{8} \operatorname{cis} 315°$.

83.

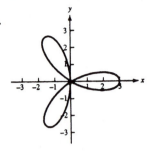

85.

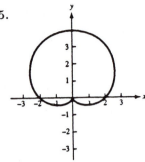

87. Use the Law of Sines. Let x be the distance between the first city and the UFO, and let y be the distance between the second city and the UFO. Let h be the height of the UFO. Add a diagonal from where the UFO is flying perpendicular to the line connecting the two cities. Then, $\dfrac{\sin 128.73^\circ}{2.3} = \dfrac{\sin 10.48^\circ}{x}$, so $x = 0.5363$. The distance from the first city is approximately 0.54 mi. Then $\dfrac{\sin 128.73^\circ}{2.3} = \dfrac{\sin 40.79^\circ}{y}$, so $y = 1.926$. The distance from the second city is approximately 1.9 mi. Using either triangle formed, $\dfrac{\sin 90^\circ}{1.926} = \dfrac{\sin 10.486^\circ}{h}$, so $h = 0.3503$. Thus, the height of the UFO is approximately 0.35 mi.

CHAPTER 7   LOGARITHMIC FUNCTIONS

Problem Set 7.1

5. a. $\frac{3^5}{3^2} = 3^{5-2} = 3^3 = 27$

 b. $\frac{4^3}{4^2} = 4^{3-2} = 4^1 = 4$

 c. $\frac{2^5}{2} = 2^{5-1} = 2^4 = 16$

7. a. $\frac{2^7}{2^{10}} = 2^{7-10} = 2^{-3} = \frac{1}{8}$

 b. $\frac{3^4}{3^7} = 3^{4-7} = 3^{-3} = \frac{1}{27}$

 c. $\frac{4^5}{4^7} = 4^{5-7} = 4^{-2} = \frac{1}{16}$

9. a. $(5^2)^{-1} = 5^{-2} = \frac{1}{25}$

 b. $(3^4)^{-1} = 3^{-4} = \frac{1}{81}$

 c. $(2^4)^{-1} = 2^{-4} = \frac{1}{16}$

11. a. $27^{\frac{2}{3}} = (3^3)^{\frac{2}{3}} = 3^2 = 9$

 b. $81^{\frac{3}{2}} = (9^2)^{\frac{3}{2}} = 9^3 = 729$

 c. $25^{\frac{3}{2}} = (5^2)^{\frac{3}{2}} = 5^3 = 125$

13. a. $e^3 = 20.08555$ b. $e^{-2} = 0.13534$

 c. $e^{0.12} = 1.12750$ d. $e^{0.08} = 1.08329$

15. a. $\left(1 + \frac{1}{1000}\right)^{1000} = 2.71692$

 b. $\left(1 + \frac{1}{100000}\right)^{100000} = 2.71828$

17. a.   b.

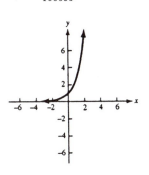   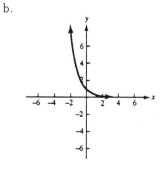

19. a. 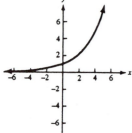   b.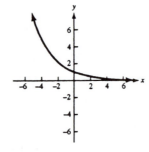

21. a. 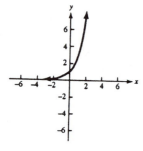   b.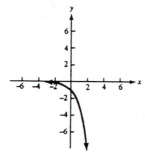

23. 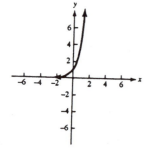   **25.**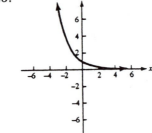

27. 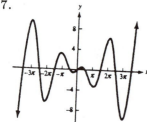   **29.**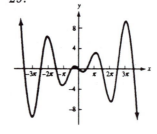

31.

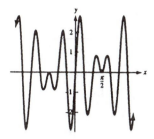

33.

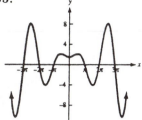

35.

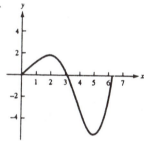

37.

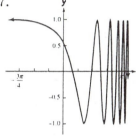

39.

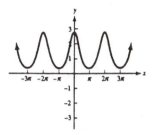

41. a.   harmonic motion

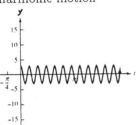

b.   resonance

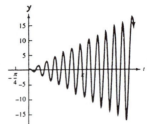

c.   damped harmonic motion

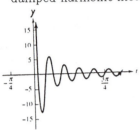

43.

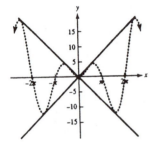

45.

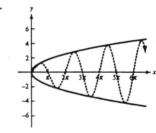

47.

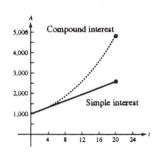

49. $3^{\sqrt{2}} \approx 4.7$

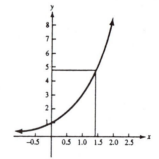

51.

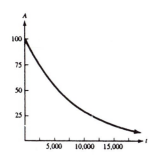

53. $P = 34,560 - 2,560 = 32,000$ and since we are given r and n, $i = \frac{r}{n} = \frac{0.12}{48}$ and $N = 48$. Thus, $m = \frac{(32,000)(0.12/48)}{1-(1+(0.12/48))^{-48}}$ and $m = \$ 708.30$.

55. $P = 215,000$ and since we are given r and n, $i = \frac{0.084}{360}$ and $N = 360$. Thus, $m = \frac{(215,000)(0.084/360)}{1-(1+(0.084/360))^{-360}}$ and $m = \$ 622.73$.

57.

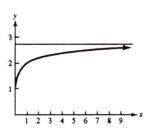

59. a. $\pi^{\sqrt{3}}$ is larger   **b.** $\pi^{\sqrt{5}}$ is larger   **c.** $(\sqrt{6})^x$ is larger

   d. When $N = 3$, $5\pi^{\sqrt{N}}$ is larger, but when $N = 6$, $(\sqrt{N})^\pi$ is larger. So this would be the case until $N > \pi^2$, then the former expression would be larger.

61. a.

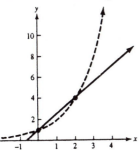

   b. The secant line passing through $(0,1)$ and $(2,4)$ is shown in part (a), the slope is 1.5.

c.

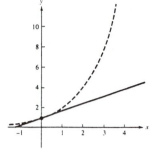

d.   The slope of the tangent line from the table in part (c) is 0.69315.

Problem Set 7.2

1.   For positive b and A, $b \neq 1$, $x = \log_b A$ so $b^x = A$ where x is the logarithm and A is the argument.

3.   The natural logarithm is a logarithm to the base e and is denoted $\ln x$.

5. $\ln N$ means $\log_e N$

9.   The three types of exponential equations are common log, natural log and arbitrary base.

11.   The growth/decay formula is $A = A_0 e^{rt}$

13.   a. $\log_5 5$ which is equivalent to $5^x = 5$, so $x = 1$

   b. $\log_5 25$ which is equivalent to $5^x = 25$, so $x = 2$

   c. $\log_5 5^{-3}$ which is equivalent to $5^x = 5^{-3}$, so $x = -3$

15.   a. $\log_b b$ which is equivalent to $b^x = b$, so $x = 1$

   b. $\log_b b$ which is equivalent to $b^x = b^3$, so $x = 3$

   c. $\log_b b^{-6}$ which is equivalent to $b^x = b^{-6}$, so $x = -6$

17.   a. $x = \log_4 8$

   b. $x = \log_5 10$

   c. $x = \log_6 4.5$

19.   a. $x = \ln 6$

   b. $x = \ln 1.8$

   c. $x = \ln 34.2$

21.   a. $3 = \log_1 01000$

   b. $2 = \log_9 81$

   c. $-1 = \log \frac{1}{e}$

23. a. $\log 4.27 \approx 0.63$

   b. $\log_b b^2 = 2$

   c. $\log_t t^3 = 3$

25. a. $\log 71600 \approx 4.85$

   b. $\log_3 9 = 2$

   c. $\log_1 91 = 0$

27. a. $\ln 2.27 \approx 0.82$

   b. $\ln 16.77 \approx 2.82$

   c. $\ln 7.3 \approx 1.99$

29. a. $\log_\pi 100 = \frac{\log 100}{\log \pi} \approx 4.02$

   b. $\log_{1.08} 5450 = \frac{\log 5450}{\log 1.08} \approx 111.79$

   c. $\log_{\sqrt 2} 8.5 = \frac{\log 8.5}{\log \sqrt 2} \approx 6.17$

31. a. $x = \frac{\ln 0.0056}{\ln 2} \approx -7.48$

   b. $x = \frac{\ln 105}{\ln 8.3} \approx 2.20$

   c. $x = \frac{\ln e^2}{\ln 10} \approx 0.87$

   d. $x = \frac{\ln e^8}{\ln 10} \approx 3.47$

33. a. $x = \log_{125} 25 = \frac{\log 25}{\log 125} = \frac{2}{3}$

   b. $x = \log_{216} 36 = \frac{\log 36}{\log 216} = \frac{2}{3}$

35. a. $x = \log_{512} 2 = \frac{\log 2}{\log 512} = \frac{1}{9}$

   b. $x = \log_{4096} 2 = \frac{\log 2}{\log 4096} = \frac{1}{12}$

37. a. $6^{5x-3} = 5$, $5x - 3 = \log_6 5$, and $x = \frac{\log_6 5 + 3}{5} = 0.07796488803$

   b. $5^{3x-1} = 0.45$, $5x - 1 = \log_5 0.45$, and $x = \frac{\log_5 0.45 + 1}{3} = 0.1679530909$

39. a. $10^{5-3x} = 0.041$, $5 - 3x = \log_1 00.041$, and $x = \frac{\log_{10} 0.041 - 5}{-3} = 2.129072048$

   b. $10^{2x-1} = 515$, $2x - 1 = \log_1 0515$, and $x = \frac{\log_{10} 515 + 1}{2} = 1.855903615$

41. a. $5^{-x} = 8$ or $\frac{1}{5}^x = 8$ and $x = \log_{1/5} 8 = -1.292029674$

   b. $7^{-x} = 125$ or $\frac{1}{7}^x = 125$ and $x = \log_{1/7} 125 = -2.481262426$

43. $3 \cdot 5^x + 30 = 105$ and $3 \cdot 5^x = 75$ so $x = 2$

45. $8\pi^x = 112$ and $\pi^x = 14$ so $x = \log_\pi 14 = 2.305397424$

47. $e^{12x \ln(1+(0.055/12))} = 2$ and $\ln(e^{0.0548743421x}) = \ln 2$ so $x = 12.63153513$

49. $x = 0.53$, $x = 3.18$

51. $x = 2.48$

53. $x = 1.5$

55. $P = 14.7e^{-0.21a}$ so $10.2 = 14.7^{-0.21a}$. Then $\frac{10.2}{14.7} = e^{-0.21a}$ and $\ln\frac{10.2}{14.7} = \ln e^{-0.21a}$ then $\ln\frac{10.2}{14.7} = -0.21a$ so $a = 1.74$ mi and converting to feet gives $a \approx 9190$ ft.

57. $P = 50e^{-t/250}$ and $10 = 50e^{-t/250}$ so $\ln\frac{10}{50} = \frac{-t}{250}$ then solving for t gives $t = (-1)(250)(\ln\frac{10}{50})$ so $t = 402$ days.

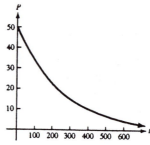

59. Using $A = A_0 e^{rt}$ gives $450000 = 209693e^{4r}$ and solving for r gives $\ln\frac{450000}{209693} = 4r$ and $r = 0.1909007566$. Then for $t = 10$, $A = 209693e^{0.1909007566(10)}$ so $A = 1414671$.

61. $A = A_0 e^{rt}$, then $50 = 100e^{2.6r}$ and $\ln 2 = 2.6r$ and $r = .26659$. Then $15.5 = 100e^{.26659t}$ and $t \approx 7$ yr.

63. $T = \frac{-E}{R\ln(1/\nu A)}$

65. $r = -\frac{1}{t}\ln\frac{I}{I_0}$

67. $h = \frac{t}{\log_{1/2}(A/A_0)}$

Problem Set 7.3

1.   a. $\log_b b^x = x$

      b. $b^{\log_b x} = x$ for $x > 0$

3.   a. $\log 10^{4.2} = 4.2\log 10 = 4.2$

      b. $\ln e^3 = 3$

      c. $\log_6 6^x = x$

5. The four types of logarithmic equations are: (a) The unknown is the logarithm. (b) The unknown is the base. (c) The logarithm of the unknown is equal to a number. (d) The logarithm of the unknown is equal to the logarithm of another number.

7. true

9. false, a natural logarithm is a logarithm in the which the base is e.

11. false, divide the $\log N$ by the $\log 5$.

13. true

15. false, $\ln \frac{x}{2} = \ln x - \ln 2$.

17. false, $\log_b AB = \log_b A + \log_b B$

19. true

21. false, $\log_b A - \log_g B = \log_b \frac{A}{B}$.

23. false, $\log N$ is negative when $N < 1$.

25. 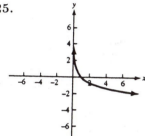   **27.**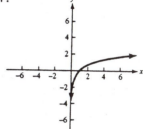

29.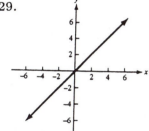

31.   a. $\log \frac{1}{10} = x, x = -1$

   b. $\log 10000 = x, x = 4$

   c. $\log 1000 = x, x = 3$

   d. $\log \frac{1}{1000} = x, x = -3$

33.   a. $\ln x = 4, x = e^4$

   b. $\ln x = \ln 14, x = 14$

   c. $\ln 9.3 = \ln x, x = 9.3$

d. $\ln 109 = \ln x, x = 109$

35. a. $3\ln 4 - 5\ln 2 + \ln 3 = \ln 4^3 - \ln 2 + \ln 3 = \ln \frac{64(3)}{32} = \ln 6$

b. $3\ln 4 - 5\ln(2+3) = \ln 4^3 - \ln 5^5 = \ln \frac{64}{3125}$

c. $3\ln 4 - 5(\ln 2 + \ln 3) = \ln 4^3 - 5\ln 2 - 5\ln 3 = \ln \frac{64}{32(243)} = \ln \frac{2}{243}$

37. a.

$$\log(x^2 - 9) - 2\log(x+3) + 3\log x = \log(x+3)(x-3) - \log(x+3)^2 + \log x^3$$
$$= \log \frac{(x+3)(x-3)x^3}{(x+3)^2} = \log \frac{x^3(x-3)}{(x+3)}$$

b. $\log(x^2 - 9) - 2[\log(x+3) + 3\log x] = \log \frac{(x+3)(x-3)}{(x+3)^2 x^6} = \log \frac{x-3}{x^6(x+3)}$

c. $\ln(x^2 - 4) - \ln(x+2) = \ln \frac{(x-2)(x+2)}{(x+2)} = \ln(x-2)$

39. a. $\frac{1}{2}x - 2 = 2$ so $x = 8$

b. $\frac{1}{2}\log x - \log 100 = 2$ and $\frac{1}{2}\log x - 2 = 2$ so $\log x = 8$ and $x = 10^8$

41. a. $\frac{1}{2}x = 5 - x$ so $x = \frac{10}{3}$

b. $\frac{1}{2}\log_b x = 3\log_b 5 - \log_b x$ and $\log_b x = 6\log_b 5 - 2\log_b x$ so $3\log_b x = 6\log_b 5$, $\log_b x = 2\log_b 5$ and $\log_b x = \log_b 5^2$ then $x = 5^2 = 25$.

43. a. $1 = x - 1$ so $x = 2$

b. $\ln e = \ln \frac{\sqrt{2}}{x} - \ln e$ and $2\ln e = \ln \frac{\sqrt{2}}{x}$, then $e^2 = \frac{\sqrt{2}}{x}$ and $\frac{e^2}{\sqrt{2}} = \frac{1}{x}$ so $x = \frac{\sqrt{2}}{e^2}$

45. a. $3 - x = 1$ and $x = 2$

b. $\ln e^3 - \ln x = 1$ and $3 - \ln x = 1$ so $x = e^2$

47. $\log(\log x) = 1$ and $\log x = 10^1$ so $x = 10^{10}$

49. $x^2 5^x = 5^x$ and $x^2 = 1$ so $x = \pm 1$

51. $\log 2 = \log 16^{1/4} - x$ and $\log 2 = \log 2 - x$ so $x = 0$

53. $\ln x + \ln(x-3) = 2$ and $x + x - 3 = e^2$ so $2x = e^2 + 3$ and $x = \frac{e^2 + 3}{2}$

55. $10^x = 4^{2x}$ and $10^x = 16^x$ since there is not an equal base then x has to equal 0 so both equal 1.

57. $\log 6^{x+2} = \log 123 + \log 10^x$ so $(x+2)\log 6 = \log 123 + x$ and $x = \dfrac{\log 123 - 2\log 6}{\log 6 - 1}$.

59. $(\log 5 - 3\log 6)x = 2\log 6 - 2\log 5$ Then $x = \dfrac{2\log 6 - 2\log 5}{\log 5 - 3\log 6}$.

61. a. $\text{pH} = -\log(.000286)$ and $\text{pH} = 3.5$

   b.   pH $= -\log(.000000631)$ and pH $= 6.2$

   c. $8.1 = -\log x$ so $x \approx 7.94 \times 10^{-9}$

63. a.   The maximum typing rate is 271 wpm.

   b.   From the equation given, $e^{-t/62.5} = 1 - \frac{N}{80}$ and $N = 80(1 - e^{-t/62.5})$.

67. Let $x = \log_b A^p$ and $y = p\log_b A$ so, $b^x = A^p$ and $b^{y/P} = A$. Then $b^x = A^p$ and $b^y = A^p$ thus, $\log_b A^p = p\log_b A$.

Problem Set 7.4

1. $\sinh 2 = \frac{e^2 - e^{-2}}{2} = 3.6269$

3. $\tanh(-1) = \frac{e^{-1} - e^1}{e^{-1} + e^1} = -0.7616$

5. $\cosh(1.2) = \frac{e^{1.2} + e^{-1.2}}{2} = 1.8107$

7. $\sinh(-2.3) = \frac{e^{-2.3} + e^{2.3}}{2} = -4.9370$

9. $\tanh(-3) = \frac{e^{-3} - e^3}{e^{-3} + e^3} = -0.9951$

11. $\cosh(-\pi) = \frac{e^{-\pi} + e^{\pi}}{2} = 11.5910$

13. $\operatorname{sech} 1.5 = \frac{1}{\cosh 1.5} = 0.4251$

15. $\coth 3\pi = \frac{1}{\tanh 3\pi} = 1.0000$

17. $\sinh^{-1} 2.5 = 1.6472$

19. $\cosh(\ln 3) = 1.6667$

21. $\tanh(\ln 3) = 0.8000$

23. $\operatorname{csch}^{-1} 0.3 = 1.9189$

25. The graph is rising for all values of x.

27.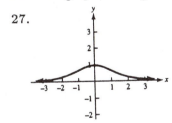

29.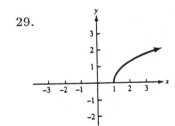

31.

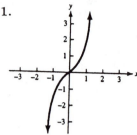

33.

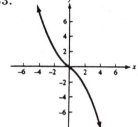

35.

$$\tanh x = \frac{\sinh x}{\cosh x}$$

$$= \frac{(e^x - e^{-x})/2}{(e^x + e^{-x})/2}$$

$$= \frac{e^x - e^{-x}}{e^x + e^{-x}}$$

37. $\dfrac{1}{\sinh} = \dfrac{1}{(e^x - e^{-x}/2)} = \dfrac{2}{e^x - e^{-x}}$

39.

$$\coth = \frac{1}{\tanh} = \frac{1/(e^x - e^{-x})}{e^x + e^{-x}}$$

$$= \frac{e^x + e^{-x}}{e^x - e^{-x}} = \frac{e^x + 1/e^x}{e^x - 1/e^x}$$

$$= \frac{(e^{2x} + 1)/e^x}{(e^{2x} - 1)/e^x} = \frac{e^{2x} + 1}{e^{2x} - 1}$$

41.

$$1 - \tanh^2 x = \left(\frac{\sinh x}{\cosh x}\right)^2$$

$$= 1 - \frac{\sinh^2 x}{\cosh^2 x} = \frac{\cosh^2 x - \sinh^2 x}{\cosh^2 x}$$

$$= \frac{1}{\cosh^2 x}$$

$$= \operatorname{sech}^2 x$$

43. $\sinh(-x) = \dfrac{e^{-x} - e^x}{2} = -\dfrac{e^x - e^{-x}}{2} = -\sinh x$

45. $\tanh(-x) = \dfrac{\sinh(-x)}{\cosh(-x)} = \dfrac{-\sinh x}{\cosh x} = -\tanh x$

47.

$$
\begin{aligned}
\sinh x \cosh y + \cosh x \sinh y &= \frac{e^x - e^{-x}}{2} \cdot \frac{e^y + e^{-y}}{2} + \frac{e^x + e^{-x}}{2} \cdot \frac{e^y - e^{-y}}{2} \\
&= \frac{e^x e^y - e^{-x} e^{-y}}{4} + \frac{e^x e^y - e^{-x} e^{-y}}{4} \\
&= \frac{2(e^x e^y - e^{-x} e^{-y})}{4} \\
&= \frac{e^{x+y} - e^{-(x+y)}}{2} \\
&= \sinh(x + y)
\end{aligned}
$$

49.

$$
\begin{aligned}
\tanh(x + y) &= \frac{\sinh(x + y)}{\cosh(x + y)} \\
&= \frac{\sinh x \cosh y + \cosh x \sinh y}{\cosh x \cosh y + \sinh x \sinh y} \\
&= \frac{(\sinh x \cosh y / \cosh x \cosh y) + (\cosh x \sinh y / \cosh x \cosh y)}{1 + (\sinh x \sinh y / \cosh x \cosh y)} \\
&= \frac{\tanh x + \tanh y}{1 + \tanh x \tanh y}
\end{aligned}
$$

51. $\cosh 2x = \cosh(x + x) = \cosh x \cosh x + \sinh x \sinh x = \cosh^2 x + \sinh^2 x$

53. $y = \frac{a}{2}(e^{x/a} + e^{-x/a})$ so $a = 40$, thus $y = 20(e^{x/40} + e^{-x/40})$. So, the poles are 40 ft apart and the height of the poles is 45 ft.

55. From Problem 54, we estimate $b = 35$, and calculate (for $a = 46$) the sag to be about 13 ft (13.97011575 ft).

57.

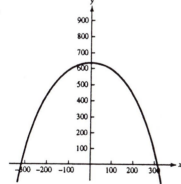

59. Let $y = \cosh^{-1} x$. Then $\cosh y = x$ and $y \geq 0$, so $\sinh y = \pm\sqrt{\cosh^2 y - 1} = \pm\sqrt{1 + x^2}$. Thus, $e^y = \cosh y + \sinh y = x \pm \sqrt{x^2 - 1}$ which is equivalent to $y = \ln(x \pm \sqrt{x^2 - 1})$.

61. Let $x = \text{sech}\, y$. Then

$$\text{sech}^{-1} x = y = \ln e^y = \ln(\cosh y + \sinh y)$$
$$= \ln\left(\frac{1 + \tanh y}{\text{sech}\, y}\right)$$
$$= \ln\left(\frac{1 + \sqrt{1 - \text{sech}^2 y}}{\text{sech}\, y}\right) = \ln\left(\frac{1 + \sqrt{1 - x^2}}{x}\right)$$

63. Let $x = \coth y$ then

$$\coth^{-1} x = y = \frac{1}{2}\ln(e^{2y}) = \frac{1}{2}\ln\frac{e^y}{e^{-y}}$$
$$= \frac{1}{2}\ln\left|\frac{\cosh y + \sinh y}{\cosh y - \sinh y}\right| = \frac{1}{2}\ln\left|\frac{\coth y + 1}{\coth y - 1}\right| = \frac{1}{2}\ln\left|\frac{x + 1}{x - 1}\right|$$

Chapter 7 Sample Test

1. a.

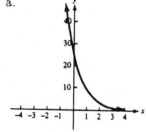

b.

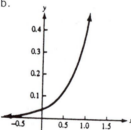

3. a.

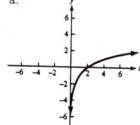

b.

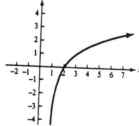

5. $\log 100 + \log \sqrt{10} + 10^{\log 0.5} = 2 + \frac{1}{2} + \frac{1}{2} = 3$

7. $\log_8 4 + \log_8 16 + \log_8 8^{2.3} = \frac{2}{3} + \frac{4}{3} + 2.3 = 4.3$

9. $e^{3x+1} = 45$ and $3x + 1 = \ln 45$ so $x = \frac{\ln 45 - 1}{3} \approx 0.935541633$

11. $\log_6 x = 4$ and $x = 6^4 = 1296$

13. $\log(x+1) = 2 + \log(x-1)$ and $\log\left(\frac{x+1}{x-1}\right) = 2$. Then $\frac{x+1}{x-1} = 100$ and solving for $x = 1.02020202$.

15. Use $b^x = A$, where $x = \log_b A$, so $A = \frac{A}{P}$, $b = (1+i)$. $A = P(1+i)^x$ and $x = \log_{1+i} \frac{A}{P}$.

17. a. $A = P(1 + \frac{r}{n})^{nt}$ and $1250 = 1000(1 + (0.07/12))^{12t}$. Then $1.25 = (1 + (0.07/12))^{12t}$ and $t = \log_{1.072290081} 1.25$ so $t \approx 3.19$ thus t is about 3 years and 2 months.

  b. $2000 = 1000(1 + \frac{0.07}{12})^{12t}$ and $2 = (1.072290081)^t$ so $t \approx 9.93$ thus t is about 9 years and 11 months.

  c. $2000 = 1000(1 + (r/12))^{(12)5}$ and $2^{1/60} = 1 + \frac{r}{12}$ so $r = 12(2^{1/60} - 1) \approx 13.9\%$.

19. a. $f(\pi) = \tanh \pi \approx 0.9962720762$

  b.

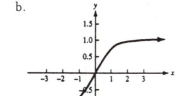

Chapter 1-7 Miscellaneous Problems

1. $\dfrac{e^{x+h} - e^x}{h}$

3. $\dfrac{5^{x+h} - 5^x}{h}$

5. $\dfrac{\ln(x+h) - \ln x}{h}$

7. a. $\log_\pi \frac{1}{\pi} = -1$

  b. $\log_2 8 = 3$

9. a. $\ln 2 = 0.69$

    b. $\ln 0.125 = -2.08$

11. a. $\cosh 1.56 = 2.48$

      b. $\sinh 1.34 = 1.78$

13. a. $\tan^{-1} 0.5 = 0.46$

      b. $\tanh^{-1} 0.5 = 0.55$

15. 1.73

17. 0.50

19. a. $\mathbb{R}$ b. $y \geq 1$ c. 1   d. $(0, 1)$

21. $y = 5 \cos \frac{\pi}{2} x$

23. $y = 6 \cos \frac{\pi}{2}(x + 1)$

25. $y = 2 \sec \frac{\pi}{4} x$

27. $y = 3 \tan \frac{\pi}{6} x$

29. $y = 3 \cos \frac{\pi}{2}(x - 4)$

31. four leaf rose

33. circle

35. cardioid

37. three leaf rose

39. lemniscate

41. three leaf rose

43. three leaf rose

45. cardioid

47. $\log 5 = \log x + \log(x + 4)$ and $\log 5 = \log x(x + 4)$ and $0 = x^2 + 4x - 5$ which factors to $0 = (x + 5)(x - 1)$ and $x = -5$ and $x = 1$. Since -5 is not in the domain, $x = 1$ is the solution.

49. $\sin 2x - e^x \sin 2x = 0$ and $\sin 2x(1 - e^x) = 0$ then $x = \frac{k\pi}{2}$ for all k contained in the integers.

51. $3 \log 3 - \frac{1}{2} \log 3 = \log \sqrt{x}$ and $\log \frac{3^3}{3^{1/2}} = \frac{1}{2} \log x$ then $1.192807137 = \frac{1}{2} \log x$ and $x = 243$.

53. $100 = 6.4(10)^{0.005x^2}$ and $15.625 = 10^{0.005x^2}$ then $\log_1 015.625 = 0.005x^2$ and $238.7640012 = x^2$ so $x = \pm 15.45$.

55. $100 = 6.4(4)^{0.005x^2}$ and $\log_4 15.625 = 0.005x^2$ then $396.5784 = x^2$ so $x = \pm 19.91$.

57. $\ln(x-1)x = 1 = \ln e$ and $(x-1)x = e$ so solving the quadratic equation for $x = \frac{1}{2} + \frac{\sqrt{1+4e}}{2} \approx 2.22$.

59.

$$
\begin{aligned}
\frac{10^{\log(\cos x)}}{10^{\log(1-\sin x)}} &= \frac{\cos x}{1-\sin x} \\
&= \frac{\cos x}{1-\sin x} \cdot \frac{\cos x}{\cos x} \\
&= \frac{\cos^2 x}{\cos x(1-\sin x)} = \frac{1-\sin^2 x}{\cos x(1-\sin x)} \\
&= \frac{1+\sin x}{\cos x} = \frac{1}{\cos x} + \frac{\sin x}{\cos x} \\
&= \sec x + \tan x
\end{aligned}
$$

61. $2 - \sin^2 3\theta = 2 - (1 - \cos^2 3\theta) = 2 - 1 + \cos^2 3\theta = 1 + \cos^2 3\theta$

63. $(\tan 5\theta - 1)(\tan 5\theta + 1) = \tan^2 5\theta - 1 = \sec^2 5\theta - 1 - 1 = \sec^2 5\theta - 2$

65.

$$
\begin{aligned}
(\sin\gamma - \cos\gamma)^2 &= \sin^2\gamma - 2\cos\gamma\sin\gamma + \cos^2\gamma \\
&= 1 - \cos^2\gamma - 2\sin\gamma\cos\gamma + \cos^2\gamma \\
&= 1 - 2\sin\gamma\cos\gamma
\end{aligned}
$$

67.

$$
\begin{aligned}
(\sec 2\theta + \csc 2\theta)^2 &= \sec^2 2\theta + 2\sec 2\theta \csc 2\theta + \csc^2 2\theta \\
&= \frac{1}{\cos^2 2\theta} + \frac{2}{\cos 2\theta \sin 2\theta} + \frac{1}{\sin^2 2\theta} \\
&= \frac{\sin^2 2\theta + 2\sin 2\theta \cos 2\theta + \cos^2 2\theta}{\cos^2 2\theta \sin^2 2\theta} \\
&= \frac{1 + 2\sin 2\theta \cos 2\theta}{\cos^2 2\theta \sin^2 2\theta}
\end{aligned}
$$

69. $\dfrac{(4-u^2)^2}{u^4}$, where $u = 2\sin\theta$. Thus,

$$
\begin{aligned}
\frac{(4 - 4\sin^2\theta)^2}{(2\sin\theta)^4} &= \frac{[4(1-\sin^2\theta)]^4}{16\sin^4\theta} \\
&= \frac{16\cos^4\theta}{16\sin^4\theta} = \cot^4\theta
\end{aligned}
$$

71. $\dfrac{u^2}{u^2+9}$, where $u = 3\tan\theta$. Thus,

$$\frac{(3\tan\theta)^2}{(3\tan\theta)^2+9} = \frac{9\tan^2\theta}{9\tan^2\theta+9}$$
$$= \frac{\tan^2\theta}{\tan^2\theta+1}$$
$$= \frac{\tan^2\theta}{\sec^2\theta}$$
$$= \frac{\sin^2\theta}{\cos^2\theta}(\cos^2\theta)$$
$$= \sin^2\theta$$

73. $\dfrac{1}{u\sqrt{u^2-1}}$, where $u = \sec\theta$. Thus,

$$\frac{1}{\sec\theta\sqrt{\sec^2\theta-1}} = \frac{1}{\sec\theta\sqrt{tan^2\theta}}$$
$$= \frac{1}{\sec\theta\tan\theta}$$
$$= \cos\theta\cot\theta$$

75. a. $\sqrt{7}\operatorname{cis}\dfrac{11\pi}{12}$, $\sqrt{7}\operatorname{cis}\dfrac{23\pi}{12}$ b. $\dfrac{\sqrt{3}}{2}+\dfrac{1}{2}i$, $-\dfrac{\sqrt{3}}{2}+\dfrac{1}{2}i$, $-i$

77. $1+\sqrt{3}i$, $-1+\sqrt{3}i$, $-1-\sqrt{3}i$, $1-\sqrt{3}i$

79. resonance

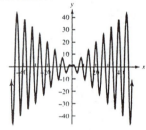

81. harmonic motion

83. harmonic motion

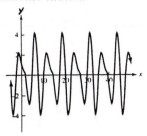

85. $\frac{1}{2}A = A_0 e^{26.5r}$ and $-ln2 = \ln\left(\frac{1}{2}\right) = 26.5r$ and $r = -\frac{\ln 2}{26.5} = .026156$. Then the equation is $30 = 55 e^{-.026156t}$ and $\ln\left(\frac{30}{55}\right) = -.026156t$ then $t = 23.173434$ h.

87.   a. $S = 6(4.5)(1.5) + \frac{3}{2}(1.5)^2 \left(\frac{\sqrt{3}-\cos 32}{\sin 32}\right) \approx 46.13$ in.

      b. $20.25 = 6(2.5)(0.75) + \frac{3}{2}^2 \left(\frac{\sqrt{3}-\cos \theta}{\sin \theta}\right)$, thus $\frac{32}{3} = \sqrt{3} - \cos \theta$ and $\theta \approx 3.9°$